Einstein in Verse

Introduction to
Special and General Relativity

By Feliciano T Bantilan, Jr

To my wife, Cynthia, and to our three sons
Hans, Kurt, and Niels
For their love and support

Contents

Preface

Books on Einstein and his theories abound. However, this book is uniquely different. It presents key concepts in Special and General Relativity, in *verse form*. The aim is to make Einstein's insights *more "fun" to learn*. It uses rhyme and rhythm to render reading memorable and thus pleasurable. In addition, what is pleasurable may foster a better understanding, as well as retention, of ideas.

Humans have a natural affinity for rhymes. Our brains perk up when we hear similar sounds of words. Rhymes get our attention and sustain it. Equally, rhymes facilitate retention of ideas, and reinforce it by repetition via a re-reading of passages again and again. Attention and retention are essential attributes in a learning situation. Musicality of the sound of language or rhythm is another feature of verse, which further strengthens already mentioned essential attributes. Rhyme and rhythm make reading enjoyable.

Use of verse apparently worked effectively in ancient times: in the Iliad and Odyssey of Homer among the Greeks; in the Vedas and Upanishads of ancient India; both rhyme, in the form of alliteration, and rhythm in Beowulf among the Anglo-Saxons, etc.

The target reader is college-educated, or college student in third year, or anyone willing to "puzzle" it out, who desires to understand the key concepts in Special and General Relativity. For example, why time slows down and lengths contract when objects are in relative motion? How Einstein-Minkowski fused space and time into spacetime? Why do all objects move at the speed of light? The book also attempts to show how Science, Einstein's theories in particular, can contribute to answering the perennial question: WHAT IT IS TO BE HUMAN; etc.

For the Arts and Humanities student, who has a liking for the big ideas from Science, which in complementation with those from the Arts and Humanities, give shape to Culture and Civilization—this may be the book you have been looking for.

Relativity has a reputation of being "difficult". For example, "curvature of spacetime" seems so abstruse, so forbidding a concept. Nevertheless, do not disarm yourself prematurely. Do not be intimidated. As it turns out, curvature of spacetime is tidal gravity—that causes the familiar ocean tides. You may even have a "gut" feel for it.

The insights of someone like Einstein are what I call the "illuminations" on the human condition by intellectuals from the "other" Culture. Whether we like it, or not; whether we are aware, or not— the reality behind these insights modulates our being in the World, as much as the reality underpinning Shakespeare's insights into human motivations, is doing. The insights of the likes of Einstein and Shakespeare are complementary in individuals with an integrated view on life. C P Snow once remarked, those in Letters and in Sciences have ceased communicating to each other, to the impoverishment of both. This book is a small bridge to connect the divide.

Didactic aim has been a part in the tradition of Poetry. The poet who readily comes to mind is Alexander Pope, as exemplified in his "Essay on Criticism" and "Essay on Man". I took inspiration from the said poet.

This book will not teach you how to solve problems in Relativity. Nor will it teach you how to prove "The shortest distance takes the longest time". No, we will spend our time grasping Einstein's insights, their implications on Reality and on the mind; and amid our quest, on what it means to be human.

More "fun" to learn does not mean that verse form is the "lazy" road to learning. It does not mean that verse form makes the difficult easy, or, the rough, plain. No; the difficult remains difficult; the rough remains rough. And to grasp it, you have to exert a determined, sustained effort and be willing to stretch your mind to accommodate the "wild" notions in Relativity. The idea motivating this book is to make the "stretching" more enjoyable relative to prose by using the rhyme-rhythm features of verse. But, "stretch" your mind, you have to. *There is no "royal, poetic" road to learning!*

More than a century has passed, since the publication of Einstein's paper on Special Relativity. Einstein's is the most recognized face in the World. Yet, his ideas have yet to find recognition in the public understanding and inform the public outlook. His insights, not only serve as a basis for practical applications, as well as theoretical investigations pushing the frontiers of knowledge, from quarks to Multi-Verse; but also can serve as a source of "illuminations" on the human condition, complementary to those originating from the Arts.

We have yet to succeed in this regard. This should give us pause to take a close look at the ways we convey Science to the public.

Have we neglected to convey to the public the humanizing aspects of Science? Have we explored enough the ways of conveying knowledge? What can the ancients teach us in this regard? What features in our long history as "homo sapiens" lend themselves to our concerns in conveying knowledge?

I suggest there is one feature in our history as humans, that stands out in view of our concerns—a feature that the ancients aptly used. It is the *oral tradition*. Since the first humans appeared, oral tradition was the sole means in transmitting knowledge for a very long time, indeed. If we fit the whole time since the first humans lived into a year, then writing started only about the morning of 30th of December.[1] That is a huge time in which oral tradition operated, i.e. about 99.5% of our time as humans.

What does this imply? We transmitted information *orally*; and we received information *aurally*! This "*oral-aural*" conveyance was the way for all information, including that of knowledge.

What further implications can we draw? During the five hundred thousand years or so, oral tradition honed our brains to receive knowledge "aurally". To me, this implies that our brains have a natural "*deep resonance*" to features of language in the oral tradition. The key language features are the rhymes and rhythms of verse.

I wrote this book in verse, inspired by this thought: to make the most out of the "resonant structures" in our brains engendered by oral tradition, to convey the deep insights of Einstein on Reality.

---

[1] The idea expressed in the statement comes from the book, "A Way with Words IV: Understanding Poetry", by Professor Michael D. C. Drout. (Drout, 2009)

It is my sincere wish that Einstein's ideas will find recognition in the public understanding and thus inform the public outlook.

I invite you, everyone:

Hop in with me, a time machine we ride,
Intent on chasing space-time concepts wild;
To fathom Einstein's insights into Reality,
In his Special and General Relativity.

Send-off Lyric for my Book:

Go, my book, go; to Publics you go;
May your form to their form mirror true;
In anxiety, I send you off to fly,
In what is surely not a clear sky;
I await the Publics' verdict for guidance;
Until then, I hold my judgement in abeyance.

Plan of Book

The whole book is divided into two books: Book I, "Space, Time and Special Relativity"; and Book II, "Physics, Gravity and Reality". Each book is further subdivided into parts.

Book I discusses the central concept of interval, in the context of the metric equation, together with supporting concepts of simultaneity, two foundational postulates, consequences coming from the metric equation such as time dilation, twin paradox, time travel, and length contraction. Book I opens with a historical context, introduces the significance of Einstein's insights on space and time; then, it outlines the arguments in the debate between absolutist versus relationist on the nature of space. It moves on to alert the reader to the obstacles to understanding Relativity—our intuitive notions, the lens with which we view the world, sculpted over eons in an environment of stability and slow motion. Book I closes with a discussion of implications and impacts from Special Relativity: epistemological, ontological, psychological, misconceptions and take-home ideas.

Book II explains two main theories on gravity: Newton's Force Theory of Universal Gravitation; and Einstein's Curvature of Spacetime for Tidal Gravity, popularly known as General Relativity. As earlier remarked, Curvature in Geometry of Spacetime may seem so abstruse, so forbidding a concept. I repeat you should not pre-maturely disarm yourself. Such concepts should not intimidate you. Book II starts with "something" out there as the object of Science, to the fragility of the basis of human knowledge, and on to the scientific method, where conjectures are the source of our theories. It goes on briefly to indicate the experimental support for General Relativity.

Book II ends with a discussion on the implications raised by the two theories of gravity in relation to reality and mind. First, we contrast and compare Newton and Einstein's theories, describing the same reality but using two completely different "pictures". Then, we consider the question: who is right, Newton or Einstein, along the way delineating the concept of domain of validity. Last, we touch on pyramid of knowledge; no Authority in Science to confer what is right.

The whole book closes with the perennial question: *what it is to be human, from the perspective of Science.*

Suggestions in reading the book: For what it is worth, I suggest the following. *First*, read Book I; then read through Book II. If you will, first enjoy the sounds: the rhymes and the rhythms, as you note the sense, which readily comes. Later, in subsequent readings, focus on the sense, even as you continue to enjoy the sound. Still later, you may fully enjoy the sound and the sense at the same time!

In the best of the ***oral-aural*** tradition, the ideal is to read the passages aloud.

In verse, conciseness or economy of words is a "rule". Thus, expositions or arguments in verse are, by design, more "transparent" than in prose.

Second, avail yourself of the notes provided, as you see fit. The notes are of two kinds, indicated by a superscript: footnotes, by superscript 1, 2, 3...; and endnotes, by i, ii, iii... Footnotes are there to help clarify terms; they are usually short. Endnotes are there for those who want to go into details of an argument or a derivation of an equation, for those who are so inclined. They also give the citations.

Third, review the properties of a right triangle in the context of the Pythagorean Theorem. The background assumed is education up to 2nd year college. Some familiarity with concepts involved in dealing with equations, like algebraic manipulation, would help. Very helpful, too, is a general understanding of the relationship of two sides and hypotenuse in a right triangle.

ENJOY!

Acknowledgment

Cynthia Bantilan, my wife, has been there for me with her love and support. She was a quiet source of strength, during the "darkest days" in my struggle with Parkinson's. In those months of groping in the "dark" as to what to follow, she had the fine instinct to let me go my way. In hindsight, it was the right decision.

Hans Bantilan read the entire manuscript in details more than once. He provided a check on the physics of my statements, as well as suggestions to improve the presentation. Kurt Bantilan volunteers to promote the book. Niels Bantilan designed the beautiful cover. I am very pleased and grateful for their support.

To my sister, Sr Librada Bantilan, MIC, who is a nun, I am thankful for her prayers and moral support during my days of trial. Just the thought of her helped ease my anxieties.

My way led me to the website of Dr Amy Yasko. I owe my "coming" back to life to Dr Amy Yasko and her staff. A year and some months into her protocol for neurological illness, I started to get back some of my "brain" and something more. Suddenly, I began to think in verse. I like reading and reciting poetry since I was young. However, I never composed a poem in my life before my partial recovery. At age sixty-five, I began to write poetry, over a year ago! Among the first poems I composed, are the two poems that make up the content of this book.

My thanks go to Jim Ryan, who read the first draft of the manuscript and provided valuable feedback in the form of questions. I am grateful too for his words of encouragement after he learned of my decision to publish the two poems.

For his encouraging comments and suggestion, I thank Uttam. He was the first person to receive a copy of the poems, outside the family. For her very positive comments, I am grateful to Padmaja.

Here, I go beyond the usual acknowledgment. I am most grateful to the small molecule, CH_3, the methyl molecule that is setting my whole methylation cycles going again!

Einstein in Verse

Introduction to
Special and General Relativity

Introduction

Hop in with me, a time machine we ride,
Intent on chasing space-time concepts wild,
To fathom Einstein's insights into Reality,
In his Special and General Relativity.

Like Wordsworth, a lonely cloud wandering,[2]
Through space and time, we will be winging,
Not to appreciate the "daffodils" of Nature;
But to understand Nature's Architecture.

We start with Special Relativity;
We'll meet the strangest thing in memory;
It is the **constancy of the speed of light;**
Scientists puzzled with all their might.

The fact is contrary to our expectations;
Scientists tried all sorts of explanations;
For such a strange behaviour,
Unsatisfactory were all conjectures.

---

[2] William Wordsworth (1770 - 1850) was a famous Romantic poet. He
wrote the poem "Daffodils".

The constancy of the speed of light,
Is not an ordinary fact, in hindsight;
Indeed, it is an intrinsic, primary behaviour;
Unexplainable by other features of Nature.

Common sense, which includes us all,
Treats this fact of light to be as usual;
Instead, with "uncommon" sense, Einstein saw
The constancy of speed of light is Nature's Law.

From this, Einstein deduced strange consequences;
On elements, like time and space, in our experiences:
The key components in design of Cosmos,
The topics in Book I, we will discuss.

We cannot get over this feeling strange;
In mind, we yield; yet intuition, unmoved, remains;
A dog can never comprehend Quantum Mechanics;[3]
It's hard: our intuitive understanding has its limits.

We then move on to General Relativity;
Einstein's theory on the phenomenon of gravity;
For ten years after Special Relativity, he struggled;
Between confidence and misgivings, he wobbled.

A key idea inspired his imagination: letting go in gravity;
One banishes gravity, by yielding to it completely!
Einstein says, it is his life's happiest thought;
The field equation was the culmination of his effort.

---

[3] Quantum Mechanics is a branch of Physics; its principles govern the behavior of atoms, molecules and elementary particles.

Say, we jump out a window; we are "weightless";
However, Tidal Gravity remains; as we will see, this is
What Einstein eventually realized: the Spacetime
Curvature; on this, in Book II, we will spend our time.

We'll be discussing implications and impacts too: such as,
What do the two theories, reveal about the Cosmos?
Facts of Nature and theories: what is their connection?
What do the theories imply, on the mind-Nature relation?

With the map of what's ahead laid out,
A ready mind and a heart so stout,
And each a seat in the time machine,
It is time for our journey to begin...

Book I: Space, Time and Special Relativity

Special Relativity, with four other papers,
Einstein published in 1905: his miracle year!
Each contributing to change the face of physics,
At a time of a "difficult" marriage, with Mileva Maric.

Part 1: Introduction

Einstein the child was slow in learning how to speak;
Rebellious he was to authority, often tongue-in-cheek;
It led a headmaster to take a drastic action, his expulsion;
Another declared he'd never amount to much, in exasperation.

Disdaining authority, he questioned common wisdom;
Such notions as space and time, in years to come;
A compass gift did stir an early sense of wonder;
His mind tended to think in terms of pictures.

A recent book calls out Einstein is wrong on time;
A long road spans from call to a new theory of time;
Einstein's insights on time and space hold true, till then;
After more than a hundred years since.

1.1 Historical Context

To get a historical sense, we step thru the rungs of ideas,
Arising from our past explorations of our Cosmos,
Our home; each rung reveals a cosmic Architecture;
Until in 1905, Einstein found a clearer blueprint of Nature.

From the ancients to 1905, the year of Special Relativity,
The Architecture of Cosmos has been an object of study;
Tying the various conceptions is the underlying question:
For any object, *what is its natural state of motion?*

Mechanical Phenomena

Come to think of it, Science essentially is a study of motion;
From quarks, everyday objects, to the largest aggregations;[4]
We figure out their trajectories thru time and space;
No motion, no change, no life; everything freezes.

Ancients-Aristotle's answer splits, according to two realms;[5]
A different set of rules governs each; for the earthly realm,
Natural state was to be at rest, as close to Cosmos' Centre;
Law of motion was: whatever moves, is moved by another.

For the heavenly realm: of moon, planets, sun, and stars;
It was motion in perfect circles: continuous, uniform, circular;
Cosmos' Centre was Earth, surrounded by rings concentric,
Carrying moon, planets, sun; outer sphere ferrying stars fixed.

It was a split-level, two-tiered Architecture;
An object's motion derived from its essence: earth, water,
Air and fire in earthly realm; the aether, the quintessence,
In heavenly realm: the perfect vis-a-vis the imperfect essence.

---

[4] Quarks are the constituents of particles like a proton. One down
quark and two up quarks make a proton. Quarks have fractional
charges. For example, down quark has -1/3 charge; up quark has +2/3
charge. So, for a proton, total charge is: $-1/3 +2/3 +2/3 = -1/3 + 4/3 = 3/3 = 1$.
[5] Aristotle (384 BC, 322 BC) was a Greek philosopher and a polymath; a
student of Plato and teacher of Alexander the Great. *Wikipedia.*

Claudius Ptolemy continued the fixation on perfect circles;[6]
Building on Aristotle's Cosmos, he set up a system of epicycles,
Circles upon circles, as many as needed to make his model fit,
The seeming wayward wanderings, of then known planets.[7]

A HUGE SHIFT resulted from idea of Nicolas Copernicus:[8]
He placed the Sun at the Centre of the Cosmos;
He relegated the Earth to a sideshow; many were displeased;
It was not Earth, but hearts and minds that he displaced.

With Copernicus, moving circles on circles planets continued;
His Sun-centred cosmology, to planets' apparent retrograde
Motions, gives a simpler explanation: the seeming stop,
Reverse and forward again, are all due to Earth's motion.

Kepler broke circle spell; on Brahe's data, 3 laws made out:[9]
Equal areas in equal times, sun-planet line sweeps out;
Not circles as ancients thought, but ellipses planet orbits are;
Cube of orbit radius is proportional, to orbit Period Square.[10]

---

[6] Claudius Ptolemy (90 AD, 168 AD) was a Greco-Roman writer of
Alexandria, known as a mathematician, astronomer and astrologer.
Wikipedia.
[7] The ancients observed that planets seem to stop, reverse its motion,
and then move forward again.
[8] Nicolas Copernicus (1473, 1543) was a Polish Renaissance
mathematician and astronomer who formulated the heliocentric
model of the Universe which placed the Sun, instead of Earth, at the
Centre. *Wikipedia.*
[9] Johannes Kepler (1571, 1630) was a German mathematician,
astronomer and astrologer; a key figure in the Scientific Revolution.
Tycho Brahe (1546, 1601) was a Danish nobleman known for his
accurate and comprehensive astronomical and planetary
observations. *Wikipedia.*
[10] The third law of Kepler relates the planetary orbit radii and orbit
periods (the time for one revolution around the sun). It says that the

To throw away perfect circles, Kepler was the first;
Yet, heavenly and earthly motion: a dichotomy did persist;
But, idea of "perfect" heavenly bodies began to break apart;
Galileo's findings fuelled the two-realm unification start.

Additional jolt to Earth-centred View Galileo delivered;[11]
Four moons orbiting Jupiter, with telescope he discovered;
Earth was not at all the centre; moreover, he saw "blemishes",
Dark spots in "perfect" Sun; "perfect" then seen tarnished.[12]

The natural state of motion for any object, is uniform motion,
Galileo says; it means motion at constant speed and direction;
We have difficulty accepting this idea, because our intuition
Continues to operate, assuming Aristotle's law of motion:

Whatever moves, is moved by another;
In Galileo's view, uniform motion needs no mover;
Mover is only needed to change state: speed and/or direction;
Inertia is tendency of objects to be at rest, or in uniform motion.

If you see an object change motion, then a push or a pull did it;
If an object is in uniform motion, then no forces act on it;
Galileo's inertia principle is a pillar, in the cosmic Architecture;
It played a key role, in the build-up of a more accurate picture.

---

ratio of the cube of the orbit radius and the square of the orbit period of any two planets is equal to a constant.

[11] Galileo (1563, 1642) was an Italian physicist, mathematician, astronomer and philosopher who played a key role in the Scientific Revolution. *Wikipedia.*

[12] The dark spots Galileo saw with his telescope are what are now called sunspots. Sunspots are temporary phenomena on the sun's surface that appear visibly dark compared to its surrounding regions. They are caused by intense magnetic activity. *Wikipedia.*

The pace of Scientific Revolution quickened with Newton;[13]
He picked up from where Galileo left off; to extend
Principle of inertia, Newton formulated his 2nd law of motion;
Force is equal to mass multiplied by acceleration.

If no force acts, the acceleration of an object is zero;
Thus, there is no change of motion; this is just the law
Of inertia; Galileo's principle of inertia, is a special instance,
In the application of the second law of Newton.

On the particular force, gravity, Newton formulated a theory;
Force of one on another equals product of masses, inversely
Proportional, to the square of the distance between them;
Universal: it applies to bodies on Earth and in the Heavens.

With his three laws of motion and universal law of gravitation,
Newton effected Earth and the Heavens Grand Unification!
Distinction was gone: perfect-imperfect bodies, 2-tiered picture;
By the 18th century, our home was a single, unified Architecture!

In Mechanics, the study of motion, is the key question:
In what frame or frames, hold the laws of motion?[14]
Laws of motion hold in uniformly moving frames;
They do not hold in accelerating frames.

Above is the *Galilean principle of relativity;*
Same Physics laws hold in frames, moving uniformly;
These are called inertial frames: all are equivalent;
There is no preferred frame of reference.

---

[13] Isaac Newton (1642, 1727) was an English physicist and
mathematician, a key figure in the Scientific Revolution. *Wikipedia.*
[14] Frame here means a "place for observing", like in the moving train
or on the stationary platform. Frame will be discussed in section 4.3,
Book I.

In everyday language, principle of relativity says:
A statement like "I am moving" is meaningless;
It implies motion is absolute; not relative;
"I am moving with respect to something", instead.

Frame moving at 600 million miles/hour, with respect to you,
Is equivalent to a frame moving at 0 speed, with respect to you;
The same laws of motion hold true, in both frames;
And neither can claim, it is the preferred frame.

Every time you ride in a plane, you experience above as true;
When you pour coffee, coffee falls inside the cup; not to you;
Except during take-off, landing or turbulence, everything,
As in your room, occurs in plane, if doing the same things.[15]

At this point, in the early part of the 18th century,
The laws of motion and the universal law of gravity,
Were well established; their swath of applications was wide;
Their truth and applicability were unquestioned, until 1905.

1.2 Significance of Special Relativity

Relativity is a sleuthing work superb of imagination;
In uncovering Nature's hidden goings-on;
It was a problem seen by common sense;
A "solution" instead, by Einstein's uncommon one.

Fewer notions if any, are so fundamental,
Of the fabric of reality, they are essential;
Than Space and Time, which everyone knows;
Change them and you change the Cosmos.

---

[15] Your room to a very good approximation is an inertial frame. It has a small acceleration with respect to sun, as it spins and revolves.

Absolute time and space are set in our brains;
These deeply ingrained notions Einstein changed;
He broke the two-century spell of the absolute;
The change was a giant step in human thought.

Fewer men if any, have changed our vision;
Our thinking since the dawn of civilization;
Than now iconic, then iconoclastic young Einstein;
A change, its public understanding we have yet to attain.

Fewer joys if any, which life affords in time so fleeting,
Which last, satisfy, and of us humans aptly becoming,
Than the joy of understanding space and time deep,
A treasure, neither moth nor rust destroys we keep.

1.3 Nature of Space

1.3.1 Absolutist versus Relationist

The stage where cosmic motion happened was space;
Time, thought unrelated to it, took a backseat;
There are no marks in space to use as meter stick,
We measure motion with respect to other objects.

In daily living, since our earliest ancestors,
We regard space as no more than a container;
There was no pressure from environment to change,
As ancestors passed on to progenies their genes.

In the 17th century, a debate raged on the nature of space;
One side, the absolutists; the other, the relationists;
Space is a thing, one says; other, only a relation among things;
Both sides agreed, on the order the space notion brings.

1.3.2 Newton's Rotating Bucket Argument

From idle talk to disputation, Newton raised the debate;
He posed the famous experiment with a rotating bucket;
Water filled and tied to rope twisted to maximum torsion,
When released, bucket spins with rope's unwinding motion.

Initially, bucket rotates but water remains stationary;
Then, water catches up with bucket and goes rotary;
Rim higher centre lower, a concave surface formation,
An absolutist or a relationist must give an explanation.

Momentum continues and twists the rope oppositely;
At maximum torsion, bucket stops momentarily;
Water continues to spin in original direction,
Surface remains concave with water's rotary motion.

At constant velocities, you cannot tell if you're in motion;
To other objects, you have to refer to get an indication;
But you know you are accelerating without an external;
Accelerating with respect to what remains essential.

Water in bucket rotating is rotating with respect to what?
With respect to bucket itself, an answer that satisfies us not:
At start, bucket rotating water stationary, flat is the surface;
At end, bucket stationary water rotating, concave is the surface.

In both situations, bucket and water have relative motion;
In one, it's flat; in other, concave: this can't be the explanation;
Newton vehemently opposed it and disagreed;
With respect to space itself, bucket water rotates, he said.

1.3.3 Newton Asserts Existence of Absolute Space

His most important discovery Newton claimed himself:
The water is spinning with respect to absolute space itself;
Water at rest with respect to absolute space: flat is the surface;
Water spins with respect to absolute space: concave is surface.

Relationists, like Leibnitz, took a beating; a retreat did ensue;
From then on, the pendulum swung to the absolutist view;
To the idea of absolute space, it was reinforcement:
Newton's reputation and his rotating bucket argument.

Newton left absolute space, vague, unclear;
Advance was made by Ernst Mach, a philosopher:
He claimed the whole Cosmos is involved in any motion;
This helped Einstein's General Relativity formulation.

From Newton's view, every time a bend you're making,
With respect to absolute space you are accelerating;
It is the ultimate reference of all motions imaginable,
But neither was it directly used nor was it measurable.

For two centuries, Newton's absolute space, time held sway;
His approach influenced all manner of scientific inquiry;
In 1905, a seismic tremor shook the pyramid of knowledge,
When Einstein's space and time relativity took centre stage.

Part 2: Obstacles to Understanding: Intuitions

Honed over eons are our instincts and intuitions;
We cannot help it: as in the Muller-Lyer illusion,[16]
But see two arrows of obviously different lengths;
Even though we know, their lengths are the same.

2.1 Source of Difficulty in Understanding Relativity

Our brains evolved in stable environment of slow motion;
This sculpted corresponding instincts and intuitions;
These are the lenses with which the world we see;
It's the source of difficulty in understanding Relativity.

Essentially, we are "stuck" in one frame of reference,[17]
Planet Earth all our lives; basically there is no difference,
Between neighbour's plane speed ride and our own 0 speed;
We're snails compared to 670 million miles/hour, light speed.

Due to snail speed, we don't have to take into account,
Our neighbour's space measure; or clock rate count;
Our sense of even flow of time, permanence of space,
In our daily living, dominates our consciousness.

---

[16] In the Muller-Lyer illusion, you are presented two straight lines of the same length. At both ends of each line are appended short segments. In one, the appended segments fold inward, thus making both ends look like a pointed arrow. In the other, the appended segments fold outward, thus making both ends look like "Y". Your brain sees the pointed-arrow line shorter than the "Y"-ending line.
[17] Reference frame will be discussed in section 4.3, Book I. Briefly, it is a coordinate system in which objects may be located in space and in time.

Again, due to snail speed, it costs us nothing,
To think of time and space as separate things;
To be aware, time is wedded to space, vice versa, is key;
This is a big source of difficulty in grasping relativity.

Thus, our common sense notions of time and space,
Literally, on so limited a range of speeds are based;
It's not surprising then: Relativity seems out of sense;
Rejection or unease is our mind's automatic defence.

Set now your mind to be aware your intuition,
Of permanent places, universal time; separation
Of time, space, will block you as you try to understand;
It keeps whispering, "it's not the way things stand".

In fact, there are no permanent places, no universal time;
There are only moving places, only a variable flow of time;
Your house is not a permanent place, though it feels so;
It's spinning, revolving; it's moving with Supercluster Virgo.

In the Cosmic view, everything is "moving";
Galaxies move relative to other galaxies;
Stars in galaxies move relative to other stars;
Planets of stars move relative to other planets.

The above view is at scales where planets are points;
We versus Earth, in turn, are tiny: we too are points;
We spin, revolve, and move together with all on it;
This fact, on our intuition and instincts, sets limits.

2.2 Need to Stretch Our Imagination

We need our imagination to work double time;
Our intuitive grasp of things and play of mind,
Are deficient: they rest on slow motion, permanent place;
Unprepared we are for subtleties of time and space.

We have to stretch our minds to bridge the gap;
Our ability to simulate scenarios mentally, we have to tap,
We need to leap our intuitive notions to Einstein's insights.
And repeat the process till we get all notions right.

Understanding key concepts of Special Relativity,
Opens our mind's reach to new vistas of reality:
Like time travel to the future, time dilation;
Relativity of simultaneity, length contraction...

Part 3: Concept of Interval: Analogue of Distance

We will use our familiarity,
With "distance" in Euclidian geometry;
Draw a straight line between two points;
However one measures it, length remains the same.

3.1 Concept of Interval

Young Einstein, then a clerk in a patent office,
Dared to question our assumptions on time and space;
Holding sway, their absoluteness as then conceived,
Both space and time, he showed, to be relative.

Minkowski, mulling over the relativity paper of Einstein,
Saw far-reaching implications: fusion of space and time;
Thus, was born the concept of *Interval:*
It is Relativity's, Special and General, Pivot-Central.

What distance is to Euclid's geometry,
Interval is to Spacetime geometry;[18]
For all, distance is the same between two points;
For all, *interval* is the same between two events.[19]

It is unusual to take time at par with space,
In defining a "distance" quantity; Minkowski, based
On results from Einstein's relativity, defined such a quantity,
The *interval*—analogue of distance in Euclid's geometry.

---

[18] Spacetime refers to the fusion of space and time. The concept will be elucidated in the book.
[19] Events are idealized occurrences marked by a point in time, its occurrence; and by a point in space, its position.

3.2 Graphs in Spacetime

In 2-dimensional Euclidean space, we have a vertical y-axis,
Perpendicular to a horizontal x-axis;
In 2-dimensional spacetime, we have a vertical t-axis,
Perpendicular to a horizontal x-axis.

In 2-dimensional spacetime, object at rest: graph is vertical line;
Object moving right: graph is straight right-tilted line;
Object moving left: graph is straight left-tilted line;
Graph of point object in spacetime we call a *world line.*

If you rotate your coordinate axes, components change,
But Euclidean distance between two points is the same;
If you move, your coordinate axes rotate in spacetime;[20]
Space and time components change, but *Interval* remains.

What could be more familiar than Time and Space?
We live these notions every day, we don't even notice;
In our habits of thought, they are so deeply ingrained,
We cannot conceive our notions have to change!

We naturally take pride in our imaginative powers,
We show them in paintings, architectures, theatres;
Yet, far greater still, are Nature's powers of imagination,
We'll see surprises in Relativity with its notions.

---

[20] The rest frame has one set of axes, t-axis and x-axis. A moving frame
has t'-axis and x'-axis rotated with respect to the t-x axes. If you move,
then for the same point or event, you get different component values
because the axes are rotated with respect to the rest frame axes. But,
the Interval is the same.

3.3 Assumption on Time and Space: Fixed

You assume Time as absolute or fixed,
Flowing equably once the ticking rate is set;
The same everywhere for everyone,
Whether clock is relatively at rest, or a moving one.

Similarly, you take Space as absolute or fixed,
The length of a meter stick is set;
The same everywhere for everyone,
Whether stick is relatively at rest, or a moving one.

I will argue you have to change your notions;
Between same two events, clocks in relative motion,
Do not measure the same time duration;
We have to give up absolute time conception.

The behaviour of sticks in relative motion,
Also exhibits a surprising length contraction;
Equally, we need to give up space as absolute;
Attaining both, we have to give a lot of thought.

Newton was wrong when he asserted these as set,
Conclusion arrived at with his experiment with bucket;
Einstein introduced the relativity of time and space;
Minkowski: quantities, which are absolute and fixed.

Part 4: Two Postulates of Special Relativity

Known long ago, first by Galileo, was principle of relativity;
What Einstein introduced was new: speed of light constancy;
He cut the Gordian knot, so to speak, by making it a postulate,
An avalanche of discoveries ensued with the opened floodgate.

4.1 Two Postulates: Foundation of Special Relativity

Space or time is relative? It's an affront to our intuitions;
But, Special Relativity rests on two solid foundations:
One, Physics laws are the same in all inertial frames;
Two, for everyone, the speed of light is the same.

The first foundation is the principle of relativity;
The second foundation is speed of light constancy.
All strange results we'll see come from these two;
It paves our path; if we're convinced the two are true.

Special Relativity: we're using a wrong title;[21]
The key concept is an absolute, the *Interval;*
Einstein-Minkowski found this quantity after much thought,
We now go through the steps to arrive at this absolute.

4.2 Choice of Units: Speed of Light is one

Let our units be such, that speed of light is one;[22]
Distance of one light-second, takes light to travel in one second;
One light-second distance, over one light-second time is one;
With time and distance in seconds, time equals distance gone.

---

[21] The title should have been "Special Theory of Invariance".
[22] In effect, we express all speeds as fractions of light speed.

4.3 Reference Frame; Inertial Frame

Reference frame is a lattice of clocks and sticks in space;[i]
At each point, a clock has six neighbour clocks, a stick apace;
Any event can be located in lattice, say a firecracker explosion,
Time is of nearest clock; count of sticks from origin, its location.

Lattice constant depends on desired precision;
All clocks in lattice undergo synchronization;
A flash is sent from origin, at zero time, to all positions;
Each clock is set equal to light travel time, to clock's location.

As described above, imagine the whole of space,
Filled with interpenetrating lattices of clocks and sticks;
Each lattice is ever moving at different constant velocity,
Each records time and space of all events, automatically.

Easier said than done, I repeat, be aware your intuition,
Of permanent places and universal time; the separation
Of time and space, will block you as you try to understand;
It keeps whispering to you, "It's not the way things stand".

I repeat, there are no permanent places, no universal time;
In truth, only moving places, only variable flow of time;
Your church is not a permanent place, though it feels so;
It is spinning, revolving; it's moving with Supercluster Virgo.

If you hop into a frame, it's natural to regard it at rest;
Large structures relative to each other, their locations are set;
It helps your understanding, if you keep above images in mind,
As we grapple through the mind-bending insights of Einstein.

Frames moving at constant velocities,
Are inertial reference frames;
In these frames, at zero net force, object's motion
Maintains the same speed and direction.

Relativity relates different frames'
Time and distance measurements;
The uniqueness of one frame,
Over others is not borne out in experiments.

Special Relativity
Seeks to arrive at an invariant quantity,
In the welter of measured times and spaces:
A "stitch" in spacetime fabric everyone agrees.

4.4 First Postulate: Principle of Relativity

"Physics laws are the same in all inertial frames" means:
Ball let go from rest, say, inside constant-velocity-flying plane,
Falls straight down, as it does, in stationary reference frame;
All inertial frames are equivalent: Physics in them is the same.

It does not matter, where you are in the Cosmos;
Be it on Mars, on planets in Andromeda, or on Uranus;
It does not matter, what constant velocity you're moving;
The same Physics laws govern whatever you are doing.

Say, two balls A, B head to each other with same speed 3 m/s;[23]
After collision: A, B reverses direction with same speed 3 m/s;
If B is at rest, A is heading right toward B at 10 m/s: now what?
Relativity principle: inertial frames are the same, answers that.

---

[23] m/s stands for meter/second.

We jump into a frame, moving right toward B at 5 m/s;
In moving frame, B moves left 5 m/s; A moves right 5 m/s;
This we know: each comes out in opposite direction at 5 m/s;
After collision: A moves left at 5 m/s, B moves right at 5 m/s.

Out of moving frame and back into rest frame, we see
A at rest, while B moves right at 10 m/s; such is a way,
To show Physics laws are the same in all inertial frames;
You get a sense what the principle of relativity means.

4.5 Second Postulate: Constancy of Speed of Light

For everyone, speed of light is the same; what does it mean?
Running after light, speed is less relative to you: common sense;
Running away from light, speed is more relative to you;
But, careful experiments show common sense notion is untrue.

4.5.1 Electric and Magnetic Phenomena

Here, we pick up where we left off the historical thread;
Theory of electric and magnetic phenomena inevitably led,
To a crossroad: the threat of a schism, at the heart of Physics;
We'll see how Einstein effected a rescue, to Physics in crisis.

Electric and magnetic phenomena were also objects of study;
Two outstanding scientists stood out: Michael Faraday,[24]
The experimentalist; James Clerk Maxwell, the theorist;[25]
Between them, they figured all phenomena electromagnetic.

---

[24] Michael Faraday (1791, 1867) was an English scientist who contributed to electromagnetism and electrochemistry. *Wikipedia.*
[25] James Clerk Maxwell (1831, 1879) was a Scottish mathematical physicist. He published his textbook, *A Treatise on Electricity and Magnetism in 1873. Wikipedia.*

4.5.2 Maxwell's Equations: Speed of Light is Constant

All of electromagnetism, Maxwell expressed in four equations,
So-called Maxwell's equations; he figured out a wave solution;
The solution comprised of electric and magnetic waves;
They propagate thru time and space at constant speed.

The constant speed he found was equal to that of light;
Light is just an electric and magnetic wave, he then realized;
So, electricity, magnetism and optics are one;
It turned out to be important: speed of light is constant.

4.5.3 Michelson-Morley Experiment

Thus, scientists regarded light as a wave;
By definition, it has to have something in which to wave;
For sound waves, medium is air; water waves, water;
So, light wave must wave in something, the aether.

Light must wave in and propagate thru its medium;
Light, from measurements, propagates at speed c in vacuum;
Aether must fill all space and possess strange properties;
It offers no resistance to bodies; has very high stiffness.

Huge speed of light implies the high stiffness of aether;
It is the reference frame, in which speed of light to measure;
If one is at rest in aether, one measures the same speed,
In any direction; medium sets the speed, no object can exceed.

If one runs east, as a bird flies east, less is speed of flight;
As in air, so aether sets the maximum speed of light;
Aether was thought to be at rest, while earth was moving;
Yet, same was light speed; any way aether wind was blowing.[26]

In 1887, Michelson and Morley measured the speed of light;
With care and precision, speed was the same to their surprise;
Like a swimmer, clocking the same speed up- or down- stream;
Or, across-back stream: this result baffled scientists no end.

Their apparatus has a viewer, a beam splitter, a source of light;
It uses wave interference: in-phase waves, image is bright;[27]
Out of phase, dark; light splits to 2 perpendicular directions;
Both reflect back to splitter, and in some combination,

Continues to enter the viewer; if earth is at rest in the aether,
And perpendicular paths are same, image is bright in viewer;
If there is a time delay in one path, viewer image is dark;
Rotating the apparatus shifts the bands, of bright and dark.

---

[26] Aether wind arises from the "supposed" relative motion of the earth and aether. The medium sets the maximum speed of any wave using the medium. A sound wave in air has a speed of about 700 miles/hour, maximum speed. If the whole air moves away from you 50 miles/hour—a 50 miles/hour wind blowing away from you—then the speed of sound from a source with respect to you is 650 miles/hour. Assuming the same logic, we expect speed of light to vary according to the direction of the light with respect to direction of aether wind.

[27] Interference is a unique property of waves. When a crest of one light wave coincides with a crest of another at a certain position, the two waves are in phase and they add: we have constructive interference—spot is bright. Similarly, the coincidence of a trough of one light wave with a trough of another at a certain position—spot is bright. But, if a crest of one light wave coincides with a trough of another at a certain position: we have destructive interference: waves are out of phase and they subtract—spot is dark. The bright and dark bands can be seen thru the viewer.

Sensitivity of their apparatus was much more than enough;
Yet, no shift of bright-dark bands was seen; aether stuff
And the mechanical thinking that underpins it, is logical
Deduction from known behaviour, of waves mechanical!

4.5.4 Crisis in Physics

In what frame is the velocity of light, equal to c?
In the frame at rest, with respect to aether, it must be;
If so, a split exists: relativity principle holds in Mechanics;[28]
A privileged frame for speed c holds in Electrodynamics.[29]

The long-standing belief in relativity principle, breaks;
If true, then a schism exists in the very soul of Physics!
Physicists, then, were cognizant of the developing crisis;
Michelson-Morley result put the crisis, scientists had to face.

4.5.4.1 Aether Drag Hypothesis

Scientists devised various ways to explain this null result;
Earth drags a volume of aether with it: a desperate thought;
Consequently, no aether wind blows in the paths of light;
A check on aether drag idea is thru the aberration of starlight.

We use an analogy to understand the aether drag idea;
You're standing in rain, holding overhead an umbrella,
As rain is pouring straight down; if you walk or run,
To stay driest, you have to tilt umbrella in run direction.

---

[20] Mechanics is a branch of Physics that deals with the description of motion; also it relates the forces acting on bodies and their motion.
[29] Electrodynamics is a branch of Physics that deals with electric and magnetic phenomena.

The faster you move, the greater the tilt in direction of motion;
Biggest change in tilt occurs, if you move in opposite directions;
Such is an analogue, of starlight aberration: one tilts telescope
To see light from stars: rain is starlight; umbrella is telescope.

Six months apart, directions of earth's motion are opposite;
One observes greatest change in tilt of telescope, as apposite;
If aether drag occurs, then there is no starlight aberration;
Let's see analogue of volume of air, dragged along in motion.

Running, you drag a big volume of air; rain outside the volume,
Falls slant toward you; raindrops enter top of still air column;
On overhead umbrella, raindrops just fall straight down;
No wind, no relative motion between you and air column.

So, it is with dragged aether volume; starlight outside volume,
Falls slant to you; starlight enters top of still aether column;
On upward telescope, starlight just falls straight down;
No wind, no relative motion between you and aether column.

But, in fact, we observed starlight aberration;
We're compelled to conclude, no aether drag in Earth's motion;
We're compelled to conclude, aether wind must be blowing;
Yet, most sensitive apparatus then, showed no bands shifting!

4.5.4.2 Lorentz-Fitzgerald Contraction Hypothesis; Lorentz Transformation

What if lengths contract in direction of motion?
Just enough, to offset the beams' time difference expectation;
Scottish Fitzgerald and Dutch Lorentz pursued this notion;
Out came the so-called, Lorentz transformation.

Originally, it related aether at rest and the moving Earth;
Aim was to make light speed constant; it turned out correct:
A transformation from one to another system of coordinates;
But, it was all ad hoc; devoid of any theory that underpins it.

Physicists, then, tried hard to understand,
This strange behaviour of light, which stood defiant;
Until a patent clerk turned the problem on its head,
From a problem to a solution: a postulate instead.

4.5.5 Einstein Resolves the Crisis

Einstein declared postulate, light speed constancy;
Together with other postulate, the principle of relativity;
Michelson-Morley experiments and Maxwell's theory agree;
He dismissed the aether; deduced consequences, aether-free.

There shows Einstein's great genius; while his peers
Saw it as a problem; instead, he saw it clear
As a starting point, to illuminate vistas he could glimpse,
From his high perch: of exciting conjectures and dreams.

4.6 Really Just one Postulate

Einstein essentially re-asserted the principle of relativity;
Both laws of motion and laws of magnetism and electricity,
Hold true in all inertial frames; there is no preferred frame,
In which light speed is c; in all frames, light speed is the same

.

Special Relativity really comes down to just one postulate;
All Physics laws are the same in all inertial frames;
This is the expanded Galilean principle of relativity;[30]
Maxwell's equations imply the speed of light constancy.

Speed of light is the same, relative to anything and everything;
No matter how fast you move, to same speed light beam clings;
We give relief to our mental unease, by realizing what is speed;
If time-space varies with motion, constant speed is guaranteed.

Take the time to understand the logic of the last line;
It'll aid your grasp of "strangeness" of space and time;
Speed is distance divided by time: if time and distance,
Vary with motion, speed of anything can be made constant.

In the postulate, it asserts that speed of light is constant;
It follows that, with motion, must vary time and distance;
As we will see in the discussion of the light clock,
Grasping relativity means, you pass this stumbling block.

---

[30] The Galilean principle of relativity referred only to the laws of motion. The expanded Galilean principle of relativity includes all laws of Physics; in particular, it includes laws expressed in Maxwell's equations. Maxwell's equations imply that the speed of light is constant.

Part 5: Relativity of Simultaneity

The concept of simultaneity is a key,
To understanding Special Relativity;
Judgement of time and distance on it depends;
Analysis of simultaneous events is where we begin.

5.1 Concept of Simultaneity

When you say, you left your house at twelve o'clock,
You meant both hands of clock were pointing up;
At the same time, you stepped out the house door.
It is a judgement based on simultaneity, at core.

Simultaneous events at each end of a stick,
Define the length of the object;
Here again, simultaneity rears its head;
This, no one, but Einstein suspected.

Events simultaneous at one place, everyone agrees;
The trouble comes for events, at distant places;
We need two synchronized clocks to know;
Two events are simultaneous or not, to show.

5.2 Simultaneity and Synchronization

Simultaneity and clock synchronization,
Are just different aspects of the same question:
Simultaneous? You need clocks in synchronicity;
Synchronous? You need to set clocks simultaneously.

5.3 Dependence of Simultaneity on Relative Motion

Chu and Sue are two inertial frames;
Sue is a shore frame; Chu is a ship frame;
Chu is moving right at speed v, according to Sue;
Sue is moving left at speed v, according to Chu.

Chu straps a firecracker at each end of the ship;
Specifically set-up, on simultaneity to fully grip;
From middle of ship, a trigger flash of light is sent;
Are two firecrackers' explosions, simultaneous events?

To Chu, he is at rest and light speed is constant;
Two firecrackers from the flash are equidistant;
Thus, the two events are simultaneous to Chu:
Ship length separates the two; but, time is zero.

In Sue's frame, Chu moves right; constant is speed of light;
Back moves to flash; front away from it: all in her sight;
Forward flash, compared to backward, has longer distance;
Thus, two explosions are not simultaneous: Sue's stance.

Events simultaneous in frame Chu;
Are not simultaneous in frame Sue;
True for any two frames in relative velocity;
This is the relativity of simultaneity!

"Tis with our judgement as our watches; none
Go just alike, yet each believes his own";[ii]
Some defect in timepieces, implied Alexander Pope;
None go just alike, due to relative motion of their own.[31]

Only space separates the two events, in frame Chu;
A mix of time and space separates the two, in frame Sue;
These results are measured quantities;
In each frame, they are objective entities.

The same conclusion obtains,
If the roles are exchanged;
Events simultaneous to Sue,
Are not simultaneous to Chu.

Result stems from light speed constancy,
And from the principle of relativity;
Difference is neither psychological,
Nor physiological; it is physical.

---

[31] The idea that relative motion gives rise to clocks measuring different times will be discussed in section 6.1.2, Book I.

Part 6: Metric Equation: Core of Special Relativity

This is the key part of Book I: the metric equation;
It defines "distance" in spacetime, for frames in relative motion;
In terms of space and time measured in different frames;
The wonder of it is: for any two events, the value is the same.

6.1 Vertical Light Clock[32]

6.1.1 Measuring Time and Space between Two Events

Our friend Chu constructs a vertical light clock;
Two mirrors L_0 apart, each to the other reflects light back;
At L_0 distance between them and light speed being one,
Each round trip takes $2L_0$ time, equal to distance gone.

Two inertial frames Sue and Chu, we're now considering;
Chu, from Sue's view, at constant velocity, v, is moving;[33]
Time, distance measured in Chu and Sue we will relate,
We will compare each frame's clock ticking rate.

Consider two events: A, a firecracker explosion;
Event B, another explosion, in another location;
Adjust L_0 such that beam start and event A coincide;
Beam return and event B as well, as clock in frame Chu rides.

---

[32] A clock is any periodic occurrence: your heartbeat, your breathing, etc. The light clock is used because its operation is very transparent. The results apply to all processes.

[33] By the principle of relativity, from Chu's view, Sue is moving with the same speed in the opposite direction. Each views its own frame to be at rest and the other as moving—again, from the principle of relativity.

Axes aligned at 0 time, 0 distance, Sue and Chu begin;
Space from A to B is zero; both occur at Chu's spatial origin;
Same distance L_0 up and down goes the beam,
Chu measures time $2L_0$ between A and B events.

At frame Sue, she sees light go up and down diagonally;
Two clocks at A, B, show longer time difference surely;
With the speed of light being a constant,
Longer time it must be, for a longer distance.

6.1.2 Clocks in Relative Motion Measure Different Times[34]

Right off, Sue concludes time runs slower in moving clocks;
Compared to time in her two stationary clocks;[35]
This she concludes from longer diagonal distance,
Coupled with the fact the speed of light is constant.

Let (z_0, z_1) be the coordinates of Chu;
Let (x_0, x_1) be the coordinates of Sue;
Subscript 0 refers to time dimension;
Subscript 1 refers to one space dimension.

Sue's coordinates relate to Chu's by a Lorentz transformation;[iii]
Where Chu is moving at speed v along positive x_1 direction;
Coordinates of event A: (0, 0); event B: $(2L_0, 0)$, for Chu;
Event A: (0, 0); event B: $2L_0/\text{sqrt}(1-v^2) * (1, v)$, for Sue.[36]

---

[34] This section contains the essential insight into the "strangeness" of time and space. I suggest you read this section when your mind is fresh.

[35] Anticipating your unease, the same conclusions hold true from Chu's view: that Sue's clock run slower, as we will see in the next section 6.1.3. Again, the principle of relativity is at work here.

From results above, we have the values for Chu and Sue;
Distance $D = 2L_0/\text{sqrt}(1-v^2) * v$ for Sue; and zero for Chu;
Time $= 2L_0/\text{sqrt}(1-v^2)$ is longer in stationary clocks, in Sue frame;
While time $= 2L_0$ is shorter for clock of Chu, in moving frame.

Measurement of time in each frame is not the same:
One clock moves from A to B to measure time, in Chu frame;
Two clocks one at A, the other at B, measure time for Sue;
The two clocks A and B are in synchrony, in frame Sue.

If you find the conclusion "strange",
Think of the two postulates again;
Sue and Chu measurements are equally valid,
Constant light speed implies time depends on relative speed.

This is where you leap over your intuition, to a new insight;
It's based on accepting the constancy of speed of light;
Or, the laws of Physics are the same in all inertial frames;
It includes electromagnetic laws of Clerk Maxwell, James.[37]

6.1.3 Escape from Inconsistency: Relativity of Simultaneity

Will this not lead to inconsistency, with principle of relativity?
If Sue concludes, that clock of Chu runs more slowly,
Then, surely, Chu will conclude the clocks of Sue run faster;
No, from his analysis, Chu concludes clocks of Sue run slower.

---

[36] Sqrt $(1-v^2)$ we call the shrink factor. It is apt because its value is equal to or less than one. Speed v is expressed as a fraction of speed of light. As v increases from 0 to 1, shrink factor decreases from 1 to 0.
[37] The laws of electricity and magnetism are contained in the four equations of Maxwell. The equations necessarily imply that the speed of light is a constant. So, accepting that all laws of Physics are the same in all inertial frames means that one accepts the constancy of speed of light in all inertial frames.

How can this possibly be?
We escape inconsistency, by the relativity of simultaneity;
Sue correctly claims, clocks A and B are synchronous;
To Chu, correctly too, clocks A and B are not synchronous.

Clock A of Sue, Chu finds does lag behind clock B,[iv]
By distance D of Sue, multiplied by velocity v;
At start, Chu's clock and Sue's A-clock both read zero;
From event A to event B, with velocity v moves Chu.

On reaching B, Chu finds clock B does read $2L_0$/shrink factor;
His time $2L_0$ equals B-clock's time, multiplied by shrink factor;
With A-clock lagging behind B-clock in starting time,
From B-clock's reading, Chu subtracts delay in time.

With time delay subtracted, A-clock's time is $2L_0$*shrink-factor;
Thus, Chu concludes that clocks of Sue are running slower;
Relativity's consistency hinges on key concept of simultaneity;
The heart of clock comparisons is clock synchronicity.

Time dilation relation between stationary and moving clock:
Time in moving clock = (shrink factor)*(time in stationary clock);
This is exactly the relation, when Sue views Chu as moving;[38]
The same relation holds true, when Chu views Sue as moving.[39]

---

[38] Case where Chu moves with respect to Sue: time of Chu = shrink factor * time of Sue (time of stationary clock) = sqrt $(1-v^2)$ * $2L_0$/sqrt $(1-v^2)$ = $2L_0$

[39] Case where Sue moves with respect to Chu: time of Sue = shrink factor * time of Chu (time in stationary clock) = sqrt $(1-v^2)$ * $2L_0$.

6.1.4 Relative Motion: Currency Converter

Let's pause a moment on "strangeness" to reflect;
Perfect watches don't have one universal rate of tick;
Time flows fast and slow depending on relative motion;
Surprised we are by power of Nature's imagination!

Each object in Cosmos carries an internal clock;
From stars, galaxies, planets, humans, quarks;
Each ticking slower or faster according to relative speed;
Think Special Relativity, think of this image!

As soon as an object moves relative to the rest,
Its clock tick rate differs from the rest;
Its rate slows down depending on relative speed;
Only near light speed, can difference be perceived.

Time slows as relative speed picks up means motion,
Is the currency converter in spacetime fusion:[40]
Converts time to space; from space to time back;
Or, converts clock to stick; and from stick to clock.

6.2 Metric Equation

6.2.1 Pythagorean Theorem

Sue measures distance D, time T, between A, B events;
Since speed of light is one, time T equals path length:
Twice the hypotenuse of the triangle, with L_0 side vertical;
With half distance from A to B, D/2 side horizontal.

---

[40] We will discuss this idea of conversion between time and space in Part 6, section 6.4, Book I, when we take up the total speed equation.

Using Pythagorean Theorem and manipulation,
One arrives at the famous metric equation:
Interval square = Time square minus Distance square;[v]
This metric equation in our minds must be seared.

The metric equation is the heart of Special Relativity;
Let's get acquainted with its simplicity and beauty;
Interval is an invariant quantity for any events A, B;
Analogue of Euclid's distance in spacetime geometry.

For events A and B, there is an infinite pair,
Of time and distance squares;
Their differences yield the same interval,
Arising from all velocities possible.

Let's apply these notions to Chu and Sue frames;[41]
Time equals $2L_0$, space equals zero, in Chu frame;
Time $T = 2L_0/\text{sqrt} (1-v^2)$; space $D = 2L_0/\text{sqrt} (1-v^2) * v$, in Sue;
Difference: $(T)^2 - (D)^2 = (2L_0)^2$ in frame Sue; and also in Chu.

Time-space square difference in frame Chu is equal,[42]
To $(2L_0)^2$ in frame Sue; both have the same value of Interval;
Measuring the same two events in any inertial frame,
The interval obtained is exactly the same.

Note Sue's time T and space D are functions of velocity;
We can generate infinite T-D pairs, by varying the speed v;
Square differences of T-D pairs give same Interval square;
A hyperbola figure results, by plotting the number pairs.

---

[41] See calculations in endnote iii.
[42] See calculations showing Sue's interval equal to $(2L_0)^2$ in endnote iii.

Each pair of time-space values implies a velocity;
Each of infinity of square-pairs equals invariant quantity;
At interval's beauty and power, you now can peek;
On this is built the one, absolute Spacetime fabric.

6.2.2 Velocity Definition of Interval

There is a velocity that is unique;
The constant velocity that a clock takes,
To be present at event A and event B;
Time in said clock is **S,** the Interval quantity.

Einstein and Minkowski must have been giddy,
When they saw this quantity;
Time and space, motions mix;
But, for all the mixing, you get a value fixed.

The mixing is different for each velocity;
Yet events A, B, have a unique interval quantity;
It remains the same, however one rotates,
By motion, the spacetime coordinates.

While the value of the Interval is unique;
Times and distances are frame-specific;
Their values depend on frame's velocity**;**
They are not invariant quantities.

6.2.3 Minkowski's Spacetime

Time by itself, or space by itself
Cannot stand separately by itself;
To paraphrase the words of Minkowski,
Time or space is a mere shadow of a true entity.

Only spacetime, a combination of the two,
Will result in a quantity true;
As given in the equation for the interval,
The quantity is absolute and central.

The metric equation is familiar;
It's the Pythagorean Theorem, with a sign peculiar;
Instead of a positive sign, it has a negative sign;
One yields Euclid's geometry, the other of spacetime.

A circle is to Euclid's geometry,
As a hyperbola is to spacetime of Relativity;
In former, positive sign: shortest path takes shortest time;
In latter, negative sign: shortest path takes longest time.

Equation of a circle is given by the Pythagorean Theorem;
Take two points A and B at distance R; let point A be origin;
Then point B can be any point on circle of radius R;
There are infinite combinations of x's and y's, for same R.

Equation of a hyperbola is given by interval quantity;[43]
Take two events A, B with interval **S**; let event A origin be;
Then event B can be any point on hyperbola of interval **S**;
There are infinite events: mixes of x's and t's, for same **S**.

One radius stands for circle's infinite number of points;
One interval represents hyperbola's infinite number of events;
In one fell swoop, Einstein cleared the infinite clutter:
"Make everything as simple as possible, but not simpler".[44]

---

[43] In standard form, the equation of hyperbola: $(y - v)^2/a^2 - (x - h)^2/b^2 = 1$. Our metric equation is: $S^2 = T^2 - D^2$. Putting it in standard form, we have $(T - 0)^2/ S^2 - (D - 0)^2 / S^2 = 1$, where $v = h = 0$, and $a = b = S$.
[44] The quote is from Einstein.

6.3 Time Dilation: Twin Paradox; Time Travel

6.3.1 Time Dilation[45]

We now consider the famous twin paradox;
Better than Physics Professors and Post docs;
In 1960's, much ink flowed arguing who was younger,
Terminologies in Relativity, then, were yet unclear.

At birth, Alice zips into space at 0.96 of light speed and back;
It takes her 7 years out and 7 years return by her clock;
At sweet 14, Alice comes home and sees her twin sister Sue;
Now, from what we learn, who is older between the two?

By an algebraic manipulation of the metric equation,
We get the time "shrinkage" relation;[vi]
Shrink factor multiplied by time in clock in stationary frame,[46]
Is equal to time in clock in moving frame.

---

[45] We are stuck with what may be a confusing terminology. Time dilation (I prefer to use time shrinkage) describes this phenomenon: the slowing down of time in a moving clock compared to a stationary clock; or, the faster flow of time in a stationary clock compared to a moving clock. The use of dilation and shrinkage simply depends on which clock you focus on. If you focus on time in moving clock, time in moving clock shrinks to 1/3 compared to time in stationary clock, for example. Or, if your focus is time in stationary clock, time in stationary clock dilates to 3X compared to time in moving clock, to continue the same example. Time dilation is always observed between clocks in relative motion. So, time dilation of one necessarily means time shrinkage of the other.

[46] The shrink factor is sqrt $\{1 - v^2\}$, where v is speed as a fraction of speed of light, and sqrt stands for the mathematical operation of extracting the square-root of a number.

Time in clock in moving frame,
Divided by shrink factor equals time in stationary frame;
Shrink factor depends on square of speed,
Expressed as fraction of light speed: usually tiny, indeed.[47]

If object's relative speed thru space is 4/5 of speed of light,
Shrink factor is square root of $\{1 - (4/5)^2\}$ equals 3/5;
If 10 years have elapsed in clock in stationary frame,
Then, 3/5 of (10): 6 years elapsed in clock in moving frame.

Or, in the time dilation picture,
Time in clock in stationary frame dilates by 1/shrink factor;
Say, in clock in moving frame, elapsed time is 6 minutes;
In clock in stationary frame, it's (6)/(3/5) equals 10 minutes.

6.3.2 Twin Paradox and Time Travel

Alice and her clock are present in leave and return;
In her trip, average 0.96 of light speed her rocket burns;
Using time dilation relation, Alice, we wager,
Is sweet 14; while Sue is 50, older by 36 years.[vii]

Wow, one stays young by moving near light speed;
Alice time travels into the future 36 years ahead;
She's just 14 years old, against her twin sister at fifty;
What could be more bizarre than this; but, is it reality?

---

[47] To get an idea of the magnitudes: a car moving at 200 miles/hour, has v = 200/670,000,000 = 0.0000003. Squaring it, we get 0.00000000000009. Very tiny indeed!

Alice inches to her teens; world leaps half a century!
She steps out into the future, six years and thirty;
This, you may say, only happens in movie theatres;
Cosmic particles time travel 29X their half-life, everywhere.[48]

Thousands of experiments done throughout the world,
Daily attest to the veracity of time dilation and more;
In particle accelerators, where speeds are near speed of light,
Time dilation with relative motion is happening left and right.

In Large Hadron Collider, time dilation governs all particles;[49]
Not to have extended lifetimes many times over is impossible;
The layers of detectors have been designed to account,
For particles' internal clocks slowing down their tick rate count.

6.3.3 Paradox or Hoax

Inertial frames are equivalent by the principle of relativity;
No one frame is privileged: alluded to above, the controversy;
So, Alice reasons, by virtue of Sue's motion, Sue is younger;
And, Sue reasons, by virtue of Alice's motion, Alice is younger.

At departure and return, Alice and Sue are present;
Time in Sue's clock is a proper time between the two events;[50]
Between same events, time in Alice's clock is also proper time;
Proper time depends on path: longer path takes shorter time.

---

[48] Half-life is the time required, probabilistically, for half of unstable particles to undergo decay.

[49] Large Hadron Collider is the largest particle accelerator in the World, straddling the border between Switzerland and France.

[50] Proper time is the time registered by a clock as it moves from event A to event B, whether at constant velocity or not. Here the same clock is present at both events A and B.

The last line is counter intuitive: your intuition whispers
It should be just the opposite; you should remember,
Time is not the same as space; in Minkowski's spacetime fusion,
Negative sign keeps space and time, distinct in the combination.

If you draw Alice's and Sue's respective world line,[51]
Sue's almost a straight line up for a frame attached to sun;
While Alice's has two straight segments: outward tilted right;
Homeward tilted left; the longer path of Alice is in plain sight.

The other objection is posed by the following assertion;
All motion is relative; as valid is the reverse situation:
Sue, earth and rest of Cosmos is accelerating away from Alice;
As Alice is heading home, rest of Cosmos, direction reverses.

By so describing, it should be Sue, who is younger;
Either they are of same age, or we reject Relativity altogether;
One English physicist, who refused to accept the paradox,
Has put it starkly as could be; it is not physics, it is a hoax.

Accelerating or not is irrelevant in the paradox of twins;
What counts is the spacetime world line path length;
Not the world line bending by acceleration in spacetime;
Alice's path is longer than Sue's is; thus, takes a shorter time.

At 335 million miles/hour, by 0.866, time slows down;
Ten seconds in clock at rest, you time travel 1.34 seconds;
At 17,500 miles/hour, by 0.999987 astronaut's time shrinks;
Every 10,000 seconds, 0.10 second astronaut is time travelling.

---

[51] To a good approximation, the world line of Sue is a straight line up along time direction which is up. Alice's world line has an x-component: outward flight is straight line tilted right to reflect the time component up and horizontal component x; homeward flight is straight line tilted left to time axis which is at x = 0.

In the world of particles, time travel to future is the reality;
At our level, time travel if ever is still very far away;
The reason is simple: the speed of light is enormous!
At 670 million miles/hour, it's just so hugely outrageous!

6.4 Total Speed Equation

6.4.1 All Objects Move at Speed of Light!

A curious fact but true: all objects move at speed of light!
What? Fastest speed on earth is far from speed of light!
Well, relativity says, don't ever separate time and space;
Speed is always: speed thru time + speed thru space.

In fact, total speed equation is re-arranged time dilation;[viii]
(Speed thru time)^2 + (Speed thru space)^2 = (Speed of Light)^2;
So, our combined speed always equals that of light: amazing!
If space speed is increasing, time speed must be slowing.[52]

Why time slows down in relative motion, we now can see;
Speed change thru time and space must opposite be;
Speed squares thru space and time must sum to one,
Until Einstein, ignorant we were about this phenomenon.

6.4.2 Heartbeat of Nature is Beat of Light

The beating heart of Nature is the beat of light;
The Standard: all to beat of light are synchronized;
As soon as relative motion carries a body, its tick rate
Slows down, depending on its relative motion rate.

---

[52] With speed of light equals 1: If space speed is 3/5 of light speed, then speed thru time is sqrt $[1-(3/5)^2]$ = 4/5 of light speed. Using the total speed equation, we have 9/25 + 16/25 = 25/25 = 1.

At rest in a frame, your clock is beating at the beat of light;
You're hurtling thru time as fast as light:
One light-second time per one light-second time;
But, as you move in same frame, speed thru time declines.

At speed of light, speed thru time totally
Converts to speed thru space; the only
Objects to attain this are light photons;[53]
We have yet to detect Einstein's gravitons.[54]

Total speed equation of objects in relative motion,
Expresses space-time fusion;
Slowing down of time only surprises,
If time is taken independent of space.

6.4.3 Right Triangle Relates Time, Space and Light

For objects in relative motion, speed of light, speed thru space
And speed thru time, in a right triangle they relate:
Horizontal side, speed thru space; vertical, speed thru time;
Joining the two is hypotenuse, equal to speed of light.

Speed of light being constant, hypotenuse is fixed;
Speed thru time and space, motion is a variable mix;
Total speed is always equal to that of light;
This fact on combined speed is not well publicized.

---

[53] Photons are particles of light, always moving at the speed of light.
For all other particles, the speed of light is a "limit" speed, which they
approach but never reach.
[54] Gravitons, thought to be the particle carriers of gravity, have yet to
be detected. They are supposed to move at the speed of light.

At rest in a frame, hypotenuse coincides with vertical side:
Space speed is zero; time speed equals speed of light.
At light speed, hypotenuse coincides with horizontal side;
Speed thru time is zero; space speed equals speed of light.

At rest in a frame, motion is all thru time: clock ticks fastest;
Speed thru time equals speed of light:[55]
At light speed, time stops and motion is all thru space;
Speed thru space equals speed of light.[56]

That's why, for a body in relative motion, time decelerates;
As some time motion is diverted to motion thru space;
Similarly, with processes in our body, the faster we move,
The more time slows down and the slower we become old.

6.4.4 Cosmic Speed Limit

The speed of light is the limit speed;
It is the limit no object can exceed;
Information transmission, as well as causal propagation,
Else, effects come before causes: a contradiction.[57]

A way to understand no object can exceed light speed,
Is thru the total speed equation: it clearly sets a limit speed;
To exceed it means a totally different set of laws;
That's a possibility; unlikely though; but who knows?

---

[55] Speed of light = speed thru time = (1 light-second time) / (1 light-second time) = 1.
[56] Speed of light = speed thru space = (1 light-second distance) / (1 light-second time) = 1.
[57] To show this will require spacetime graphs of two frames moving relative to each other superposed in one grid of coordinates.

Another way of seeing why speed of light is the limit,
Is to consider an object that is pushed to exceed it;
As object is accelerated, inertia increases to infinity,[58]
By 1/shrink factor, attaining immobility when speed is unity.[59]

Whether there are particles always moving,
At speeds the limit set by light exceeding;
Is still an open question; there are hypothetical tachyons,
As of now, no trace yet of these particles has been found.

6.4.5 Cosmic "Laziness"

Speed is only thru time, if body in a frame is at rest;
Proper time is a maximum; clock ticks fastest;[60]
In flat space, the shortest path is a straight line;
In flat spacetime, the shortest path takes the longest time.[61]

Bodies move to maximize proper time between events;
Events A and B are specified by time and distance;
A body moves from A to B at constant speed and direction,
Such that it is present at A, B at specified time and location.

---

[58] Inertia is the resistance of a body to change in motion, or to acceleration. As object pushed nears the speed of light, v nears 1 and the square root of $(1-v^2)$ nears zero and inertia nears infinity.

[59] Shrink factor = sqrt $\{1 - v^2\}$, where sqrt is the square root of a number and v is speed as fraction of light speed.

[60] Proper time is the time registered by a clock as it moves from event A to event B, whether at constant velocity or not. Here the same clock is present at both events A and B.

[61] This statement assumes that an object moves from point A to point B at a given time. Given the constraints, an object moves at constant speed and direction: any deviation from its constant velocity—like slowing down and then speeding up—will mean less time overall: compensation is not possible because functions depend on the square of speed. The faster one moves, the slower the time.

In resume, for any two events,
Spacetime interval is distance between them;
Component times and distances vary with velocity;
All give the same interval, the invariant quantity.

We all are in relative motion, thru both time and space;
At light speed, time is zero; we only move thru space;
At rest in a frame, distance is zero; we only move thru time;
At moving frames, all processes experience dilation of time.

6.5 Length Contraction

6.5.1 Horizontal Light Clock

We move on to consider length contraction;
Are atoms squeezed for length diminution?
As in the FitzGerald contraction hypothesis;
Or does it have to do with measuring process?

But, first the question: what is length?
Interval of two simultaneous events, at object's ends;
For a stick at rest, the two simultaneous events,
Are coincident marks of ends on a scale: its length.

Now, consider the length of a moving object;
Again, a light clock, now horizontal, we will pick;
Lying in Chu's frame parallel to direction of motion;
Lengths measured by Sue and Chu vary as shown.

As measured by Sue, let L be the length of light clock;
From one mirror to the other, light bounces back;
In going, beam chases right mirror, thus distance longer;
In return, beam meets left mirror, thus distance shorter.

In time t_1, beam hits right mirror, covers L distance plus;
In time t_2, beam hits left mirror, covers L distance minus;
After algebraic manipulation and comparison,[ix]
We get the same shrink factor for length contraction.[62]

6.5.2 Only Length Parallel to Direction of Motion

Length of stick in moving-frame depends on speed square;
It's equal to stationary-frame length, multiplied by shrink factor;
A stick in stationary frame: measured length is longer;
A stick in moving frame: measured length is shorter.

This holds for lengths parallel to direction of motion;
Shrinkage at ordinary speed is miniscule, like time dilation;
Only when moving at a sizable fraction of light speed,[63]
Will contraction be noticeable; or spectacular, indeed!

Perpendicular to direction of motion,
There is no length contraction;
It will violate the principle of relativity;
It'll pick one frame as special among infinity.

Arrange the vertical lengths to mark each other,
As the frames, pass one another;
If marks left on the vertical lengths are unequal,
Then relativity principle is false, as one is special.

---

[62] Shrink factor = sqrt $\{1 - v^2\}$, where sqrt is the square root of a number and v is speed as fraction of light speed.
[63] Moving at 4/5 the speed of light, we have $(1 - v^2) = (1 - 16/25) = 9/25$. Taking the square root, the shrink factor is 3/5 = 0.6. A 10-foot anaconda "shrinks" to 6 feet!

6.5.3 Where Contraction Comes From

Wherefrom does length contraction come?
Ultimately, we say the two postulates it's from;
To the point, it issues from the relativity of simultaneity;
In moving frames, perfect clocks beat differently.

Two surveyors A and B are measuring a path width;
A's set-up is at an angle; B's, perpendicularly oriented;
A measures longer width; B, shorter: values are different;
Difference arises from orientation of measurement.

So, length contraction does not come from squeezing,
Of atoms; rather like the two surveyors above measuring,
Path width differently; observers in the two frames,
Agree not on simultaneity; hence, values are not the same.

Ah, relativity of simultaneity!
Relative motion begets time a-synchronicity;
What is simultaneous in one frame,
Is not simultaneous in the other frame.

Oh Relative Motion, the source of a-synchronicity;
As well, on you depends events' simultaneity;
You surprised us with teen Alice, by time dilation;
As well, by a shorter wand, by length contraction.

In the particle world, motions are near the speed of light;
Loafer particles are punished by shortened span of life;
Energetic particles are rewarded by proportional longevity;
The particle world is literally a world of a-synchronicity!

A picture is worth a thousand words, they say;
Relatively moving at 0.87 of light speed, looking out you'll see,
A smoker puffs a cigar half as long as yours, truly;
Yet, it lasts twice as long as yours does, surprisingly.

6.5.4 Stanford Linear Accelerator: a Different World!

Time dilation, in the form of the twin paradox,
Gets the lion's share of attention; here, I'll talk
About a structure in California: Stanford Linear Accelerator;
Its 2-mile length was set by Relativity, not Administrators!

To get near light speed, electrons get energy boost of 40 GeV;[64]
Per meter distance, electrons gain energy of about 13 MeV;[65]
If Cosmos is in accord with Newton's laws, in how it behaves,
In a 2-centimeter distance, an electron can attain light speed![66]

In fact, the accelerator is 2 miles long; not 2 centimeters,
It is the familiar shrink factor at work: it is ever harder
To push the electron; its inertia increases to infinity,
As shrink factor goes to zero, as speed approaches unity.

USD 300 million is the cost; relativity requires 3,000 meters,
To get near the speed of light; and not just 2 centimeters!
What a world of difference between 2 cm and 2 miles!
If you are not impressed, I do not know what will!

---

[64] GeV stands for a billion electron volts. One electron volt is 1.6×10^{-19} joule.
[65] MeV stands for a million electron volts. Length of accelerator is about 3,000 meters. So, 40 GeV / (3×10^3 meters) = (40×10^3 MeV) / (3×10^3 meters) = 13 MeV/meter.
[66] In Newtonian mechanics, kinetic energy for a particle moving at speed of light c, is ½ mc^2 = ½ (0.5 MeV) = ¼ MeV. So, ¼ MeV/(13 MeV/meter) = 1/52 meter = 2 cm.

Profound difference between Einstein's world and Newton's,
Stems from this: speed of light is the same for everyone!
This is just one small corner, in the vast Physics expanse,
Which is touched by the fact, the speed of light is constant.

For us at rest on Earth, we see a 2-mile long structure;
For the electron, it's just a 4-centimeter long accelerator![67]
Here, we have a dramatic example of length contraction!
Building hasn't collapsed; severe shrinkage is due to motion.

6.6 Light Cones and Causality Structure

6.6.1 Events and Their Light Cones

Being the limit speed, speed of light defines causality;
Events fall into categories of influence possibility;
The categories are frame-independent;
Light "cones" classify spacetime events.

Pick an event, O: at a point in time and place,
A spherical shell of light expands thru time and space;
The shell becomes a circle in two dimensions;
The flash of light goes on and on in all directions.

Let's picture this in spacetime: time, t, vertical;
Origin at O; with x, -x, y, -y axes horizontal;
Flash begins expanding at O as a point circle;
Growing, rising, tracing a cone at 45-degree angle.[68]

------

[67] The time component of the energy-momentum 4-vector is $E = m/\sqrt{1-v^2}$. This is beyond the scope of the book. But, it gives a handy way of calculating the shrink factor. We have shrink factor = $\sqrt{1-v^2} = m/E = (0.5 \text{ MeV})/(40 \times 10^3 \text{ MeV}) = 1.2 \times 10^{-5}$. The contracted length is $(3 \times 10^3)(1.2 \times 10^{-5}) = 4$ centimeters.

Thus is formed the future light cone of event O;
Its mirror image is the past light cone of event O;
It is the same process in reverse: a huge circle,
To O, contracting, rising, coming at 45-degree angle.

Past is the cone shell of light rays, converging to O;
Future is the cone shell of light rays, diverging from O;
For cones of events, light is not necessary;
Light is used just to mark off the cone boundary.

Every event in Cosmos, a light cone defines;
Light cones are everywhere and every when;
I call them the cones of influence aptly;
It is Nature's limits on causal possibility.

6.6.2 Causal Structure in Spacetime

Different events related to event O:
Events on the future light cone of O;
Events on the past light cone of O;
Events inside the future light cone of O;
Events inside the past light cone of O;
Events outside the cone of O.

Light-like related events are **on** the cone of light;
These events occur to or are set off by light;
Their time equals their space separation: interval is null;[69]
Distinct two points without a "distance", we find unusual.

---

[68] Flash spreads in all directions in two dimensions we are considering. The cone is at 45 degree angle because the speed of light is one: along the t-axis, one light-second time; along the space axis, one light-second distance—giving the ratio of one, or 45 degrees.

[69] Interval equals $(T^2 - D^2)$. For light, one light-second time is equal to one light-second distance. Hence, $(T^2 - D^2)$ is zero.

Events inside the cone, to event O, are **time-like** related;
From event O, a massive particle could be emitted,
To influence any event, **in** the future cone of O;
And O could be affected by events, **in** the past cone of O.

For two **time-like** related events, there is a frame
In which the two events occur in the same place;
However, there is no frame
In which the two events occur at the same time.

A magnetic storm in Sun occurred 8 minutes ago;
Being 8 light-minutes away, the storm can affect you[70]
Now; magnetic storm and you now can be related causally;
With speed of light the limit speed, this is as it should be.

If information could travel faster than speed of light,
Then a change in TV channels can occur long or right,
Before pressing a button; we'll be unable to distinguish,
Cause from effect; notion of causality will be extinguished!

Events outside the cone are **space-like** related to event O;
They cannot affect or be affected by O.
There is not enough time to connect the two:
The event outside the cone and event O.

For two **space-like** related events, there is a frame
In which the two events occur at the same time;
However, there is no frame
In which the two events occur in the same place.

---

[70] Light-minute is the distance light travels in a minute. Light-year is the distance light travels in one year.

A Jupiter probe lands softly on the surface;
If from now, a rock will hit the probe in 10 minutes;
Even if you have a way the damage to prevent,
Being 30 light-minutes away, you can't affect the event.

A plus-minus 30-minute-band exists, in which Jupiter events
Cannot influence or be influenced by your now event;
These events are neither in future or in past cone of your now;
Relativity gives us a glimpse of order in Cosmos we know.

Events in Jupiter in the 30-minute band, may occur with Now,
At the same time, before, or after: is there a contradiction? No!
There is no causal relation between these events and your Now;
Depending on motions of observers: three temporal orders go.

Beyond 30 minutes ago, events, which occurred in Jupiter,
Are in the past cone of your now; events that will occur,
More than 30 minutes from now, are in future cone of now;
Both sets of events can causally be related to your event now.

Such is the causal structure of our Cosmos;
The effect is expected to come after the cause;
The causal structure is invariant: independent of frame;
Else, an effect can precede its cause: chaos would reign.

Creating a big stir recently is the experimental finding,
Of non-locality; it seems a breach in the long-standing,[71]
Belief in the local character of phenomena: if something,
Is lost in a place, someone in the place must be stealing.

---

[71] Non-locality is the direct interaction between objects separated in space without intermediate mechanism or agency. Experiments on quantum particles, in the 1980's, show non-locality.

Above describes the local character of causality;
A local occurrence is produced by a local agency;
Not by action at a distance, by an agent far away;
That it violates the speed limit set by special relativity.

Debate is: was information transmitted?
If so, did it exceed the limit speed?
If true, then we have to re-think,
The whole issue of causal links.

6.6.3 Personal Light Cones

Idealized, each of us is a point in time and space;
An event in spacetime at every moment in a place;
So, all we said above about events for us hold true;
What does this causal structure mean to me or to you?

It means wherever we are, at each moment,
We define a cone, just like any event;
We can influence events, in our future cone;
We may be influenced by events, in our past cone.

Much of what is possible causally,
Is of no interest to us personally;
What is outside the cone of causality,
Is what we hanker after in our reverie.

Much as we want it, we cannot exert influence,
Faster than the speed of light; nor be influenced
By something faster than light allows;
Causality is ruled by Nature's laws.

6.7 In Relativity, NOT Everything is Relative!

Absolute space does not exist;
Absolute time does not exist;
But, Absolute Spacetime DOES EXIST;
Not Everything Is Relative! Einstein insists.

Part 7: Implications and Impacts

After going thru the labyrinths of spacetime, we now take stock:
What in Special Relativity can and have impact?
What does it say on relation between Cosmos and humans?
Does it have anything to say on what it means to be human?

7.1 Epistemological[72]

Aether gives a glimpse of the scientific process;
In dealing with wave phenomena, having had success,
To propose a propagation medium was natural, logical;
Precise experiments, though, showed result was null.

Confronted by null result, scientists of highest calibre,
Staggered, confused; unable to see through the veil;
Young Einstein, in his early twenties, saw the king "naked";
He turned the problem into a postulate, instead.

After much head scratching, aether was unnecessary;
Einstein led its dismissal and proposed a new theory;
This is the way knowledge of reality accumulates:
Conjecture, deduction, elimination: a pyramid we create.

A theory proposes; experiment disposes,
Or it may not; this epistemological loop ever continues,
Until a theory attains provisional winner-status at the top;
Cycle begins when old cracks; a new guess covers the gap.

---

[72] Of or relating to an account of acquisition and validity of knowledge.

A lesson in Epistemology: a starting point is necessary;
Facts put in by fiat, like the two postulates of relativity;
Everything, then, is arrived at by logical deduction:
Like the interval, time dilation, length contraction.

Knowledge of reality comes from our conjectures;
Created together by intuition, logic, and metaphors;
In a community critiquing ideas with evidence;
Selection is made when prediction jibes with experiment.

7.2 Ontological[73]

Interval is absolute;
Speed of light is absolute;
Four-dimensional spacetime is absolute;
Made of events whose intervals are absolute.

Spacetime exists, independently of reference frames;
In its totality are the past, present, future—all events;
Your death; Caesar crossing the Rubicon, his murder;
All that and innumerably hugely more are all in there.

7.3 Psychological

There is no privileged frame of reference;
All inertial frames are equal or equivalent;
Principle of democracy is consonant with Relativity;
This braces the resolve of those in quest for equality.

---

[73] Of or relating to the nature of existence or being.

In the physical world, motion induces observers' frames;
Measured values of the same quantities are not the same;
In the moral world, in "reference frames" we all reside;
Variedness of moral assessments is many times multiplied.

Surely, if "views" on purely physical events differ;
Legitimately based on physical reasons and no other;
Then, our views, say, on justice or poverty,
For a million different reasons are expected to vary.

For reasons psychological, physiological, and sociological,
Our views on moral matters differ: which is just natural;
The lesson I take from Relativity is summed up thus:
Strive to view from as many angles; strike a consensus.

Among the possible formulations, say, on justice;
Strive for a core formulation almost everyone agrees;
A view on any moral issue contains a truth, a grain;
To reconcile these "truths", is what we aim.

In mankind's experiment in governance;
Political and moral considerations advanced;
Then, Kings, Popes, or Chiefs ruled absolutely;
Now, Democratic Societies are ruled by majority.

Some ways evolved to protect the minority;
"Interval-like" unanimity remains an ideal entity;
Our psychological differences will not permit,
Views on moral matters completely to coalesce.

Another lesson is awareness, which is crucial;
Our views are, by habit, parochial or provincial;
As infinity of frames has the same invariant quantity,
We strive for "universality" on issues like morality.

Special Relativity re-enforces this spill over:
Each operates in a "frame", let us be aware;
Each has a take, different from the take of another;
A goal is a consensus, reconciling one to the other.

A thought bubble, we find ourselves in;
Just like observers in inertial frames;
As observers their measurements compare;
We burst our bubble by seeing bubbles of others.

7.4 Misconceptions

An alleged lesson from relativity: "Everything is relative";
Positions range from anything goes, to no truth objective;
This is absolutely nonsense Relativity interpretation;
Theory's name ought to be "Absolute Theory of Motion".

True, time by itself and space by itself are relative;
But absolute quantities take their place, instead;
The speed of light is absolute speed;
The interval, for any two events, is absolute indeed.

Another misconception: undue importance is given to observer;
Lattice of clocks and meter sticks could replace him or her;
Time, distance measured are objective truths in each frame;
But objective truth for all is the interval, true for all frames.

Time and distance are not relative, due to subjective viewpoint;
Measured values are not subjective, but objective data points;
The observer in a frame is not the source of relativity;
"Relativity" arises when relating frames in relative velocity.

7.5 Take-home Ideas

7.5.1 Physics Insights

Special Relativity can be summarised in one sentence:
Physics laws are the same, in all inertial frames of reference;
You can play tennis in Venus; use your microwave in Mercury;
Your reference frame can move with any constant velocity.

Interval, "distance" between two spacetime events,
Remains solid knowledge more than a century, since;
Time and space, relative motions dilate and shrink respectively,
And mix them to form the *interval*, an invariant quantity.

This is an image of Nature, from Special Relativity:
Each object has an internal clock; *relative motion is currency*,
Converting time to space, and space to time back;
Or, converting clock to stick, and stick to clock.

In our example, Chu's space is a mix of Sue's space and time;
In general, one's space is a mix of another's space and time;
And one's time is a mix of another's time and space;
Disagreement on simultaneity spills over to time and space.

The *standard beat* of Nature is the *beat of light*;
It's constant, no matter what: Einstein's great insight;
A right-triangle relationship: hypotenuse is speed of light;
One leg is speed thru space; other leg is speed thru time.

To remember Special Relativity, grab a meter stick;
Find a right-angled floor-wall; on them prop the stick;
Shine light on stick to cast shadows: stick is speed of light;
Floor shadow is speed thru space; wall, speed thru time.

7.5.2 Musings on Nature and Humans

7.5.2.1 Insights and Reality

Such insights into Nature, as summarized above are great;
What good are they? Reality behind these insights modulates,
Our being in the World, as much as the reality underpinning,
Shakespeare's insights into human motivations, is doing.

Let's take a bright example: the ubiquitous GPS device;[74]
Pinpoint accuracy rests on tiny warping of spacetime;[75]
The reality behind the musings on spacetime curvature,
Has spawn technology cascades, a new wave of culture.

Let's take a dark example: the mushroom cloud;
Our doom may come, if hydrogen bombs explode;
The reality behind such musings on Nature,
Is an existential threat to our Future.

Insights of the likes of Einstein or Shakespeare,
Are complementary to each other;
One makes us see our place in Nature;
The other, our place in the Social Structure.

Imagination is more important than knowledge;
Albert Einstein once said;
It is as indispensable a tool, for the scientist,
As it is usually thought so, for the artist.

---

[74] GPS, Global Positioning System, is a space-based satellite navigation system that provides location and time information in all weather conditions.
[75] Warping of spacetime will be discussed in Book II.

There are two moments in creation:
In conception and in appreciation;
Imagination in both we need,
As surely, in your experience, you appreciated.

7.5.2.2 Some Thoughts on Reality

Reality is not given once and for all;
It is reached by action, which yields little by little;
To find the perfect human image, artists are ever chiselling;
To reach reality, scientists are ever experimenting, theorizing.

Reality is not a fixed entity, like a buried treasure;
It is more like a great piece of Art, or Literature.
In poems, poets attempt to capture human's core being;
In theories, scientists try to fathom "deep down things"[x].

In a poem, a poet explores what it is to be human;
Only echoes of the poet's own being, the poet finds;
In a theory, a scientist scrutinizes physical Reality;
Also finds echoes of the scientist's own identity.

A poem is a mirror, reflecting individuality,
Reflecting humanity;
A theory is a mirror, reflecting a mind's structure,
Reflecting Nature's Architecture.

Reality speaks in the language of mathematics,
As Galileo once remarked; special relativity physics,
Reinforces this notion, of the mathematical property,
Of the relations, of the elements of Reality.

7.5.2.3 Mind and Nature

Constancy of the speed of light is a fact of Nature;
Until this fact became a part of a theory,
A brute fact it was, in its stark simplicity;
To say that facts determine theories holds no water.

It took Einstein's "uncommon" sense,
To transform it into a theory; it explained,
Other facts of Nature; more importantly,
Predicted other facts found true: confirmatory of theory.

Similitude there is, in mind's architecture,
And the architecture of Nature;
Einstein's space and time relativity,
Find their mark in Reality.

Amid our quest, we find us among the glories,
As well as the wastes, of time and space;
Subject to the same forces, binding all to the Cosmos;
Yet *different*: in thought, we're free from Nature's laws.

Mind is the atom's way, of appreciating its own,
Truth and Beauty; sentience grew over eons,
From dim groping in deeps, to clear probing of laws;
Atom has come full circle: mind mirrors Nature's laws.

For so long, only Beauty was perceived;
Humans, alas! Also Truth is now conceived!
Nature has found, its voice in us;
To search its laws, befalls on us.

To search for Truth, is a duty;
It makes us human, fully;
It is a call from Nature;
A call from us: shall we not honour?

Keats says, "Beauty is Truth; Truth, Beauty";
Artists' are ever striving for Beauty;
Scientists are ever digging for Truth;
Searching for both is Life's Total Worth.

Like Wordsworth, in vacant or in pensive mood;
The insights learned should be a light in our abode;
If daffodils in Nature can serve our inward eye,
Then, theories on Nature can fill our craving mind.

Book II: Physics, Gravity and Reality

In 1915, Einstein published his theory on gravity;
His physics was much stronger than his math ability;
The mathematics of curvature is calculus of tensors;
Delay of his field equation was due to math errors.

Part 1: Reality, Mind and the Scientific Method

Reality is as slippery as an eel;
Nets were knitted from Plato to Hegel;
You grasp it: momentarily it lingers,
But, as easily, slips thru your fingers.

1.1 Reality

Something is out there, independent of my senses;
It hurts as I stub my toe, when a cue I miss;
Whatever it is, which is out there outside of me,
That is what I mean, when I refer to reality.

Reality is like a fish, to bag it has been our quest,
Stripped to its barest, our net is our senses;
We see, we hear, we touch reality at its tips,
The rest we triangulate, using our guesses.

The lights we see at night are such tips from reality;
The data are just pinpricks of light; without theory,
We end up thinking of things, outlandish and bizarre,
And it took thousands of years, to figure they are stars.

Stars are huge balls of gas, incandescent,
Their radii are millions of kilometres, in length;
Every second, they burn millions of tons of mass,
Each gram's energy equals bomb Hiroshima's.

No one has direct experience, of the doings of stars,
Yet, by data and theory, we know the info is true by far;
Amazing, how fragile is the basis of our knowledge,
Yet, its reach is incredibly long, with pride we acknowledge.

1.2 Mind and the Scientific Method

Experience is not the source of our theory;
The source is our originality and creativity;
Experience enables us, one of our guesses to pick;
It well explains the part of reality we select.

Ultimately, our knowledge precariously rests,
On our nervous system, parsing streams of data bits;
Crackles of electrical pulses, over a network of neurons,
Far different from raw data, our senses were turned on.

Wonders of wonders, somehow we manage,
These digital neural crackles, into some knowledge,
Which explains a part of reality we are interested;
Oftentimes though, our attempts do not succeed.

Importance of data cannot be overstated,
It is the reason for the rapid progress we've made;
Pure speculations of ancients became stale;
Over millennia, they got us no nearer to the Holy Grail.

A data set in itself though, is uncertain;
We don't even know what it represents,
Until a guess we make, a concepts-net,
That forms an explanation that fits.

A data set without theory is blind;
A theory without data is lame;
The power of data and theory, when combined,
Sifts good ideas to use, while sets the bad to flame.

Good ideas are fitted to well explain,
The reality behind a tip, which captivates our brain;
The guesses are ever closely strained;
What passes through, is knowledge gained.

The steps in the scientific method are three:
One, out of your mind, you make a guess and see;
Two, deductions you draw, what the guess implies;
Three, test them to verify, if the guess flies.

The guess that flies becomes a theory;
In a few minds for a start, it holds sway;
To a growing community, it slowly percolates,
Eventually, we will know the theory's limits.

Theories do not come from data, or perspiration;
Nor, are they a product of reasoning by induction;
Nor, from the book of Nature are they read;
But conjectures which to mind, by creativity are fed.

What could be more commonplace than gravity!
The very reality, we all are in contact every day!
What could be more intriguing than gravitation!
It set the whole Cosmos into motion!

A theory is a new way of seeing our experience;
A vision of what the Universe is, and how it makes sense;
Through the eyes of Newton, as force we will view gravity;
Of Einstein, we'll view it as curvature of spacetime geometry.

This is a tale of two titans in human thought;
Illuminations on the Cosmos, humanity has sought;
Isaac Newton gave us Laws of Motion; Universal Gravitation;
Albert Einstein topped Newton, with his geometry of motion.

Part 2: Newton's Forces; Universal Law of Gravitation

Newton was not the first rationalist: cold reason;
But "he was the last of the magicians";[76]
He dabbled in the esoteric, the occult, and alchemy;
At the same time, as he probed into physics, astronomy.

2.1 Introduction

Newton towers in physical, mathematical insights; techniques;
He created new mathematics, needed in his new physics;
He invented differential, integral calculus to prove his guesses:
Like bodies over distances act as if they are point masses.

He was a man of extraordinary powers of concentration;
He could hold a problem in mind for days, till its resolution;
Truth obsessed, he didn't tolerate reasoning: false, shallow;
As master of Mint, he sent counterfeiters to the gallows.

To reveal Cosmos' secrets God destined him, Newton believed;
On optics, alchemy, theology, to gravity: notions he conceived;
But, ideas laid out in forbidding book, assured his immortality;
The book is "Mathematical Principles of Natural Philosophy".

In it, is contained a System of the World:
Every object in the Universe, on every other exerts a force,
To the tune of three laws of motion; universal law of gravity;
Thus, movements on earth and in the heavens fold into a unity.

---

[76] Lecture by John Maynard Keynes on "Newton, The Man", in
http://www-history.mcs.st-
andrews.ac.uk/Extras/Keynes_Newton.html

2.2 Newton's Force Hypothesis

Newton saw bodies in motion, but not the forces;
He saw planets revolving around the sun, but not the forces;
He saw moons orbiting about the planets, but not the forces;
The concept force of gravity was his hypothesis.

With same force, apple's and moon's accelerations he deduced,
To Inverse Square Law of Gravity, all motions were reduced;[77]
Thus, he united as one, motions on earth and in the heavens;
It was the result of Newton's triangulation, with force as given.

Three laws of Kepler from law of inverse square come out;
Equal areas in equal times, sun-planet line sweeps out;
Not circles as ancients thought, but ellipses planet orbits are;
Cube of orbit radius is proportional, to orbit Period Square.[78]

Newton's friends and foes alike, questioned the force of gravity;
How does the sun exert the force, to planets so far away?
Further, the force one body exerts on another is instantaneous;
It makes the force of gravity, mysterious and doubly suspicious.

---

[77] Inverse square law of gravitation says that force between two bodies is directly proportional to product of their masses and inversely proportional to square of distance between them. So, at twice the original distance, force becomes one fourth of original value. At three times the original distance, force becomes one ninth of original value.
[78] The third law of Kepler relates the planetary orbit radii and orbit periods (the time for one revolution around the sun). It says that the ratio of the cube of the orbit radius and the square of the orbit period of any two planets is equal to a constant.

Force of gravity till his death, Newton failed to explain,
Though successes after successes followed its train;
The nub of the problem is "action" at a distance;
There was nothing between bodies, to act as conveyance.

No question that Newton believed in God, the Creator;
Who, in Divine cipher, carefully encrypted the laws of Nature;
To pry open the secret codes, was Newton's undying quest,
In consequence, he scarcely paid attention to all the rest.

The headlong rush to apply Newton's way went on and on,
From tides, to predictions of Uranus and Neptune,
To a world system without God, this to many was a shock;
It ended in a conception of the Cosmos, ticking like a Clock.

The fish that Newton bagged is force, with other concepts;
Absolute space and absolute time, on which his system rests;
Absolute space: a container of points, with fixed locations;
Absolute time: clocks ticking equably, even in relative motion.

Absolute space matches our intuition, of a container at rest;
If anything is moving, the reference with which to test;
Absolute time jibes with our intuition, of a clock set once,
Keeps time absolutely, the same everywhere for everyone.

Gravity is a force;
Mass is its source;
It is instantaneously transmitted at a distance;
There is nothing in between for conveyance.

What now we ask, "Is Newton right"?
Is there really a gravitational force?
Is his guess right?
Right, what is its source?

Admittedly, the question requires a nuanced answer;
Yes, gravitational force still figures, and builds skyscrapers;
Yes, gravitational force still figures, and builds towers, cranes;
Yes, gravitational force still figures, in its validity domain.

No, massive bodies are beyond its scope;
No, gravitational force has hit the limits of its domain;
The curvature of spacetime with these can cope;
With it, come a slew of new concepts in its train.

Part 3: Einstein's General Relativity

Hilbert the mathematician, who bested Einstein,
In deriving the justly famous Einstein field equation,
Circulated a joke: every street schoolchild knows
The math of Relativity, which Einstein has difficulty!

3.1 Introduction

Einstein towers in physical insight;
In mathematics though, he didn't show equal might;
General Relativity is a crowning jewel, in human thought;
The culmination of Classical Physics, Einstein has brought.

After graduating, landing a job Einstein had trouble;
He showed no deference to his professors, while in school;
No professor recommended him, to any position academic;
He struggled to make ends meet, as young father wed to Maric.

Finally, he got a job in a patent office, as "inspector third class";
With steady income, he could pursue in free time, his ideas;
In spite of fatherhood: one hand rocking a cradle, other a pen,
He wrote the paper on time and space, at age six and twice ten.

In a burst of creativity, 4 papers he wrote and special relativity
In 1905; each could make anybody, a claim to posterity;
To extend his special relativity to all, constant or not, motions,
He struggled, ending in 1915 with his theory on gravitation.

This is his magnum opus;
Among the many jewels, in his works' corpus;
It is at once a testament, to Einstein's power of imagination;
And a tribute to his indomitable spirit, in pursuing gravitation.

As Bronowski once said, there are two moments in creation;
First, in the original conception; second in appreciation;
To appreciate creative works, you need to re-create the insight;
I hope you get your eureka moment, with as much delight!

3.2 Minkowski's Absolute Spacetime Fabric

In Newton's theory, time is independent of space;
Each moves in their separate, unconnected ways;
Equably from past, to present, to future, time flows;
Space spans the breadth, width, and depth of Cosmos.

Special Relativity dictates, the fusion of space and time;[79]
Each event needs a meter stick, and a clock, which times;
One to measure its occurrence time; the other its position;
Motion mixes observer's spatial distance, and time duration.

Minkowski fused space and time, by defining the interval,[80]
For any two events, yielding a value absolute: identical
For all observers, whatever their constant velocity;
Minkowski based it on Einstein's Special Relativity.

Newton gave us, absolute space and absolute time;
Minkowski, absolute four-dimensional spacetime;
Out of Minkowski's absolute spacetime fabric of reality,
Einstein fashioned his General Relativity.

---

[79] Special Relativity shows the close connection of space and time for an accurate description of the world we live in. The arena for all events or happenings in the Universe is SPACETIME. In Book I, we learn that motion mixes observer's space and time measurements.
[80] See Book I, for discussion of interval.

3.3 Concept of Curvature: of Space; of Time

If a Euclidean property does not hold,
Then, the space in which it occurs is curved;
Say, examine two parallel lines' property,
If they violate a theorem, of Euclid's geometry.

On a sheet of paper, an instance of a flat space:
Draw two parallel lines; by Euclid's postulate,
The two lines will not converge; if they converge,
Euclid's postulate doesn't hold: the space is curved.

Apply this finding to the surface of a sphere;
Draw a straight line to North Pole from equator;
Draw another straight line to North Pole;
These two lines are parallel at the equator;

But, they cross at Pole, "forced" by space curvature;
In two-dimensional space of a surface of a sphere,
Now, we know a Euclidian property doesn't hold;
This behaviour shows the space is curved.

For identical clocks, if same ticking rate doesn't hold,
Then time, a surprise to you perhaps, is curved;
Like two clocks, in an up accelerating elevator,
One at the top, and the other down the floor.[81]

---

[81] Discussion on time curvature is found in Book II, section 3.4.4.

3.4 Equivalence Principle: Physical Foundation

3.4.1 Free-Fall Thought Experiments

Young Einstein knew there was a flaw,
In Newton's formulation of gravity's law;
On distance depends, say, the force of Sun-Mercury;[xi]
Distance inclusion in a law, violates principle of relativity.

Laws of physics must be the same, in all inertial frames;[xii]
However, in all frames, distance is simply not the same;
Distance between same points, depends on frame's velocity;[82]
Something is wrong in Newton's law of gravity.

One thought experiment after another, Einstein did see,
In a free-falling elevator, everything remains force-free;
You feel weightless, as if there is no gravity;
To young Einstein's mind, this was a HUGE key.

In this Einstein thought experiment, a toy gun pops a ball;
It traces a parabola: up, then down it goes at gun level;
You and screen opaque, with a narrow horizontal slot,
At gun level, start a free fall when the ball the gun pops.

You adjust the motion of slot, to see the ball all the time;
After repetitions, to see the ball the whole flight you find,
You need to move the slot up, at constant speed; you see
The ball moving in a straight line, at constant velocity.[xiii]

---

[82] Length contraction, which depends on velocity, is one of the results of Special Relativity (See Book I).

Something magical is happening! What a thrilling discovery!
As if by magic, we banish gravity, by giving in to it wholly;
To top it all, a parabola following Newton's law, when seen,
Becomes a straight line, from a free-falling frame!

This is miraculous! A marvel of Nature's ingenuity!
Gravity vanishes not by opposing, but yielding to it completely!
And things in a free-falling frame, obey the first law of Newton!
Unless a force acts on it, object continues at rest or in motion!

A real-life example of a free-falling frame is a capsule in orbit;
It is aptly small, that earth's pull on its parts differs not a bit;
You see on TV, astronauts float as well as their food, gadgets;
When they push an object, it moves at constant speed, straight.

This insight led Einstein to expand the relativity principle:
Physics laws formulated in one frame hold true in all;
Be it an inertial frame of special relativity, in gravity's absence;
Or, a small free-falling inertial frame, in gravity's presence.[83]

A free-falling elevator, in presence of gravity,
Is like the inertial frame, in special relativity;[84]
In elevator, as in ship, doing experiments,
Results are the same:

A body at rest, or in motion,
Continues its state of rest, or motion;
Thus, a small free-falling frame
Is inertial, even in gravity's presence.

---

[83] "Small" has a technical meaning: it means space occupied by the cabin is small enough such that no variation of gravity is measurable.
[84] Special Relativity inertial frames are ones moving at constant velocity—at constant speed and constant direction.

Einstein was widening the reach of relativity principle;
He was thrilled to see laws of special relativity, hold all,
Even in presence of gravity; he was beginning to glimpse,
A formulation of a theory of motion, true for all frames.

His quest was a theory, which encompasses all motions,
Without picking one "preferred" frame; his intuition
Strongly felt, Nature does not favour one velocity;
If it were so, for the Lord, he would feel so sorry.

Let's take our bearings and pause for a moment;
The same Physics: in inertial frames, in gravity's absence;
And in free-falling frames, in gravity's presence;
With or without gravity, "inertial" means the same.

It does not matter where you are free falling;
Into the Sun, or Alpha Centauri you may be jumping;
The same Physics laws govern your motion;
As principle of relativity widens its sphere of action.

3.4.2 Equivalence of Gravity and Acceleration

At zero gravity, far from matter source,
Just like on earth, you feel a force,
In an up accelerating elevator:
It pulls you down the floor.

You feel as heavy, as if there is gravity;
Everything happens, as in situations ordinary;
You release a key, and down it falls;
His faith grew in expanded relativity principle.

You can't distinguish effects in elevator up accelerating,
From effects of gravity, while on earth's surface standing.
This in-distinguishability is the Equivalence Principle (EP),
The basis of General Relativity and his results all.

Meaning of equivalence of gravity and acceleration:
Effects of gravity can be produced by acceleration;
Effects of acceleration can be produced by gravity;
We can attribute effects of one, to the other's identity.

3.4.3 Space Curvature

A laser beam emitted parallel to the floor,
In an up accelerating elevator,
Is bent downward toward the floor:
A path with curvature.[85]

You can't tell this bent path, by EP,
From effects of gravity;
Thus, mass bends the path of light:
Warps space geometry.

---

[85] As the laser beam transits from one side to the opposite side of the
elevator, the elevator would have moved ever greater distances up at
each Pico-second. So the beam would hit a point below the point it
would hit if the elevator is at zero velocity or at constant velocity. The
beam path is curved. By the Equivalence Principle, you can't
distinguish this effect in an accelerating elevator from the effects of
gravity; then mass warps space, resulting in curvature in the geometry
of space.

Due to acceleration, path of light is bent;
The same can be attributed to gravity;
From gravity viewpoint,
Mass produces curved space geometry.

That's why, near massive bodies,
Objects follow paths, which curve;
Like planets around the sun,
Geometry tells them so to swerve.

3.4.4 Time Curvature

Consider two clocks further,
In an up accelerating elevator;
One clock T at the top, another
Identical clock D down the floor.

Compared to clock D's rate,
Clock T at the top ticks at faster rate:
T's signals arrive at floor ever earlier,
As distances are ever shorter.[xiv]

You can't tell this effect on clocks, by EP,
From effects of gravity;
Thus, mass distorts time at different places:
Warps time geometry.

3.4.5 Spacetime Curvature

Thus, Einstein's great insight is to see,
That force is not to be;
But mass warps both space and time geometries,
Giving rise to gravity.

In the acceleration picture, the path of light is bent;
At different positions, clock-ticking rates are different;
In the gravity picture, mass tells space how to curve,
And clocks to adjust their ticking rates as perturbed.

Wait a minute, you say, perfectly good timepieces,
Tick at different rates, in mass' presence at different places?
Yes, the farther from the mass, the faster is the ticking rate,
Giving the lie to the universal time, Newton did assuredly state.

For all you know, you have never seen this difference,
Our environment is such, that we live a Goldilocks' experience;
Only with truly massive bodies, the difference is significant,
Else, on earth, watches will tick the same rate each second.

Take an example of a black hole of ten solar masses;
At one centimetre above its event horizon, watches[xv]
Will tick 6 million times slower than those far from it;
One year there means millions of years' advance, for the rest!

Time travel to the future, anyone?
It looks like Physics is a lot of fun;
However, we're still pre-occupied with survival;
The limits are only set by the technological.

By yielding to gravity, we transform away "gravity",
By a suitable acceleration, can we banish all forces away?
The overall force on a body, yes; but not the tidal forces!
Of stretching and squeezing of objects, they are the causes.

3.5 Tidal Gravity is Spacetime Curvature

3.5.1 Earth's Tides

Earth is in a free-fall, to moon-earth mass centre;
It is "weightless": scale and earth both fall together;
But it "feels" forces stretching and squeezing it;
With deformable waters, giving rise to tides to the limit.

The ebb and flow of tides on earth are due to tidal gravity;
It is the "residual forces", acting on any extended body;[86]
As explained by Newton's theory, Earth's parts feel a force,
Of different strengths and directions from moon, the source.

Earth-parts-to-centre-of-moon distances
Explain the strength differences;
Earth-parts-to-centre-of-moon alignments
Account for direction differences.[87]

The pull on right is slightly leftward;
The pull on left is slightly rightward;
Together make an inward squeeze;
Far is weaker than near: make a stretch.

---

[86] Residual means the force left on the component parts of a body, when the body is in a free-falling frame. The body is "weightless"— gravity vanishes. But, because the force of gravity depends on distances of components of the body to the center point of source of force, different parts will experience different magnitudes of force, as well as different directions of force.

[87] Newton proved that objects like the moon act on other objects as if its whole mass is concentrated at a point at center. To do this, he had to invent integral calculus.

Ocean's stretch along the moon's direction
Results from strength differences;
Ocean's squeeze transverse to moon's direction,
Comes from direction differences.

Appreciate the daring of young Einstein;
He just discarded absolute space and time of Newton;
Now, he seeks to replace Newton's gravity law,
To explain the stretch-squeeze without the flaw!

3.5.2 Straight Lines in Small Free-Falling Frames

We know the meaning of curvature
In two-dimensional space;
Now, what is meant by curvature
In four-dimensional time-space?

Surprise! It's related to old familiar tides;
Heaving oceans by tidal forces: Newton's pride;
So-called "residual forces" which remain,
Even after a body is in a free falling frame.

In absence of gravity, a Special Relativity world,
Spacetime is flat, not curved;
In all inertial frames, free particles move
In straight lines at constant speed.

This property is not changed, by presence of gravity:
Free particles, subject to no forces, except gravity,
Trace "straight lines" in small frames falling-free,
Where instruments cannot detect variations in gravity.

Surveyor's "straight lines"
Are small segments of geodesics:
Great circles, which go round
The Earth's spherical surface;

So the "straight lines" of free particles above
Are segments of "geodesics", analogue
Of straight lines in Euclidian flat space;
Paths in gravity, free particles trace.

3.5.3 Kip Thorne's Thought Experiment

The moment he saw this, Einstein recognized
Tidal gravity is the curvature of spacetime;
What Newton saw as "residual forces",
Einstein saw as spacetime curvatures!

To see it, do Kip Thorne's thought experiment: stand[xvi]
At North Pole, with two balls in your hands;
Throw them up parallel; from peak, they fall,
Unimpeded through earth, and cross at centre.

Initially parallel, the two lines meet: spacetime curves!
Both balls are in free fall: gravity vanishes; but "residual" force,
Tidal force makes the two balls meet, in the view of Newton;
Spacetime curvature makes them cross, in the view of Einstein.

Thus, curvature in four-dimensional spacetime is tidal gravity;
"Residual" force left, after a body to gravity yields completely;
Every time you see a bulge, or a pinch of ocean tides,
Think, in spacetime, ocean waters are taking curvature rides!

Spacetime may seem slight, or muscle-less to us;
In fact, it is a power genie at the call of mass;
At moon's behest, it lifts tons to form the tides;
At sun's command, it holds the rails for orbs to ride.

Tidal gravity acts on all objects; spacetime geometry
Stretches and squeezes atoms, molecules in a body,
As geometry of spherical surface "squeezes" lines to meet;
This is what General Relativity is: its gist.

3.5.4 Curvature Made Visceral

We too are subject to spacetime curvature, i.e. to tidal gravity;
As usual, effects are noticeable only near value's extremity;
Near a black hole, stretching, squeezing could be so enormous;
Your body could turn to spaghetti, by colossal tidal forces.

I certainly don't feel the curvature of spacetime;
I did not feel a stretching, or squeezing at any time;
Earth's mass's too small, to produce a curvature sizable;
Sun's mass with earth's radius produces one noticeable!

Imagine you feel your head and feet are parting;
Imagine you feel your left and right sides colliding;
You know spacetime curvature you're in is appreciable;
In mind, you've just made curvature concrete, palpable!

Spacetime curvature is a stretching and a squeezing,
Along front-back, left-right, up-down as time is flowing;
If free particles' geodesics converge, curvature is positive;[88]
If free particles' geodesics diverge, curvature is negative.

---

[88] A geodesic is a path in spacetime traced by a free particle in a free-falling frame, with no force acting on it, except gravity.

Above ideas in mind, the image of ocean tides helps fix;
Stretching along moon's direction, divergence of geodesics
Makes particles, if unrestrained, go off surge,
By Earth's opposite curvature: thus forming only a bulge.

A picture of spacetime curvature: release tiny clocks free;
Paths they trace and times they show, map the geometry:
Where they converge, warp is positive; diverge, negative;
Times, depending on motion and position, show warpage.

To repeat, laws of physics are the same,
In all inertial frames;
In inertial frames of Special Relativity,
In absence of gravity;

In small free-falling frames of General Relativity,
In presence of gravity:
Frames freely falling anywhere: on Mercury,
On black holes, on the Sun, on centre of galaxy.

At this point, the physical ideas you clearly see,
The basis of Einstein's General Relativity;
Translating these physical ideas into precise math,
Success of Einstein was a daunting and lonely path.

We leave General Relativity's mathematics,
Tensor calculus, to graduate students in Physics;
You should feel good on the key ideas you learn,
They're part of your outlook, as you do the daily concerns.

3.5.5 Irony of Einstein's Joke on Minkowski

As noted earlier, Einstein fashioned his theory,
Out of the absolute spacetime fabric of Minkowski;
The irony is that, at first, Einstein was not impressed,
With spacetime fusion via interval, Minkowski addressed.

Minkowski's spacetime fusion, Einstein dismissed
As merely Relativity in a new mathematical dress;
His joke was that relativity formulation by mathematicians,
Made it so difficult, physicists no longer could understand.

Minkowski was Einstein's professor,
And called him a lazy dog; reading his papers,
However, he was impressed by his deep insights;
He was surprised: the lazy dog could bite.

The joke was actually on Einstein; he worried infinity
Of observers meant infinite warpages, of time and space;
After four years, he realized spacetime's indispensability:
It provided ready one warpage for any two events, not infinity.

3.5.6 Ontology of Absolute Spacetime Fabric

The spacetime fabric: absolute, unique, is the same
As seen in all reference frames;
The spacetime fabric: absolute, unique, exists
Independently of reference frames.

3.6 Interpretation: Einstein's Geometric View

3.6.1 Conveyor of Interaction: Spacetime Geometry

There is no force of gravity, acting on bodies at a distance;
This solved the puzzle, that troubled Newton all at once;
Curvature of spacetime geometry is the go-between,
The interaction messenger one body to another sends.

If you hold your arms level, no gravity force acts downwards;
The only force there is, is the force you yourself exert upwards,
That keeps your arms constantly accelerating away,
From the shortest path in the region's curved geometry.

The "natural" frame is one, which yields to gravity;
Like a free-falling elevator; if on bed you stay,
You're accelerating with respect to someone in free-fall;
View situations this way and strangeness of Relativity will fall.

Sun doesn't pull the planets, their orbital motions to explain,
The planets slide down chutes, in geometry the sun maintains;
No push or pull, but meanderings o'er hills, dales, and passes,
As laid out by spacetime geometry set by distribution of masses.

3.6.2 In Curved Spacetime, Shortest Path Takes Longest Time

The straight line is the shortest distance, in flat space;
In curved space, like the surface of a sphere, geodesics;
The shortest path is the straight line, in flat spacetime;
In curved spacetime, shortest path takes the longest time.

A ball thrown from A, arrives at B in ten seconds;
The path the ball actually takes, a parabola, is one,
In which a clock attached to ball shows the longest time;
It follows the curvature in geometry of spacetime.

In moving from A to B, by going up some, its clock ticks faster,
It could zip to highest point and down, if it were the only factor;
Due to time dilation of special relativity, the faster it moves,[xvii]
The slower the ticks; trade-off produces longest time curves.

3.6.3 Two Contrasting Views on Gravity

Analogy illuminates contrasting views of Newton and Einstein;
Two bodies on sphere: head north, trace shortest lines;
Paths converge: Newton saw it as effect of attractive force;
But, Einstein saw it not as effect of force; but of curvature.

There is no force of gravity! What a strange thing to realize!
The Cheshire cat in Alice: cat remains; its grin de-materializes!
Gravity remains, but force disappears: replaced by curvature;
This surely is not the last story: more surprises from Nature.

3.7 Einstein Field Equation

3.7.1 Hilbert Beat Einstein in Deriving Einstein Equation

With mass, the geometry of spacetime is linked inextricably;
Mass curves spacetime around it; and this curvature is gravity;
Space and time aren't fixed; they curve with motion, position;
Such curvatures in spacetime are traced by bodies in motion.

To express this relation, in math, between mass and spacetime,
Was the most arduous task, Einstein undertook in his life-time;
He asked a mathematician, to teach him calculus of tensors;
To express how mass-pressure densities produce curvatures.

In mathematics, Einstein was not as technically proficient;
As, for example, David Hilbert, the top mathematician then;
Errors he found one after another, delayed his final derivation;
Hilbert beat him, in arriving at famous Einstein Field Equation.

Hilbert invited Einstein, to deliver lectures on Relativity;
Einstein delivered six two-hour lectures on his theory;
Mulling over the lectures, Hilbert, with a Master's touch,
Derived the equation, five days ere Einstein could match.

To repeat, space and time are not absolute or fixed;
Time flows and space lengths are not set;
They expand, contract, curve: spacetime is rubbery,
Both motion and mass warp spacetime geometry.

3.7.2 Einstein Field Equation: Law of Spacetime Curvature

His ten-year-long struggle reached a culmination;
In the justly famous **Einstein Field Equation:**
It states, **"mass and pressure warp spacetime"**, [xviii]
It is a fitting epitaph for THE MAN FOR ALL TIME.

Symbolically, *Einstein Field Equation* can be written:

$$\begin{pmatrix} Geometry\ of \\ spacetime \\ curvature \end{pmatrix} = G \begin{pmatrix} Spacetime \\ mass\ and \\ pressure \end{pmatrix}$$

In more details, the above symbolic equation,
Where **G** is the constant of gravitation:

Choose any spacetime point location;
Pick any frame and three directions:
Left-right; up-down; and front-back;
Mass-pressure in vicinity sets spacetime warp.

Curvature pushes and pulls free particles;
It's signalled by particles' geodesics behaviour;
If geodesics are pushed together, curvature is positive;
If geodesics are pulled apart, curvature is negative.

Rate at which geodesics are pushed together,
Or pulled apart, is proportional to strength of curvature:
Along left-right, front-back, or up-down orientations;
Sum the strengths of curvatures, in the three directions.

Einstein Field Equation on spacetime curvature: [xix]
Sum of strengths of the three curvatures,
Is proportional to mass density,[89]
Plus 3 times matter pressure, in vicinity.[90]

If you and I are at the same location,
But move relative to each other;
Our space and time will differ;
Mass density, others too will differ due to motion.

I get my sum of 3 strengths of curvatures;
My mass density; my matter pressure;[xx]
You get your sum of 3 strengths of curvatures;
Your mass density; your matter pressure.

---

[89] Mass density is the mass of an object divided by the volume of the object.
[90] Pressure is force divided by area.

Whatever values we measure,
We always find the sum of the three curvatures,
Is proportional to our mass density,
Plus 3 times our matter pressure, in vicinity.

Einstein's spacetime curvature law,
Holds the principle of relativity true;
The idea of formulating laws, that are frame free,
Has become the standard, when creating a theory.

The goal Einstein set from start of his journey;
For laws true in all frames, with or without gravity;
He attained in his spacetime curvature relation;
Another way of calling it is *Einstein Field Equation*.

3.7.3 Einstein Field Equation: in Resume

In language concise, Reality seen in the New Vision,
As contained in the famous *Einstein Field Equation*:

Mass and pressure tell spacetime how to curve;[xxi]
Spacetime tells masses how to move;
In place of forces' pushes and pulls,
Is the geometry of spacetime curvatures.

The four lines above encapsulate,
The whole Cosmos' motion and dynamics;
From the very beginning, till the end of time,
Cosmos dances to Field Equation of Einstein.

3.7.4 Congratulations! We Made It!

Together, we made it to the end;
The end, alas, is where a new start begins:
Viewing Nature, not with youth's thoughtless look;
But with fascination: having glimpsed at Nature's book.

Whenever you read in the papers on black holes,
Or supernova explosion, quasars, or wormholes;
Say to yourself with a smile of understanding,
Yes, those are different types of spacetime warping.

Now you begin to see, with a new pair of eyes,
What was once inscrutable, in knowing surprise;
What before was beyond your ken,
Now becomes happily a joy to comprehend.

Part 4: Experimental Support for General Relativity

Your theory will only be as good, as its predictions are true;
Your theory may be truly beautiful; but if untrue,
Then, hard as it is, you have to discard your theory;
Distinguishing mark of scientific theory is falsifiability.

4.1 Precession of Orbit of Mercury

The orbit of Mercury, for a time has remained a mystery;
Shift excess of its perihelion, unsolved by Newton's theory;
But, curved geometry of spacetime could readily expound,
The 43 seconds of arc advance per century, of its perihelion.[xxii]

The excess advance of the perihelion of Mercury,
Is due to an additional term, arising from General Relativity;
This may be seen as additional force, beyond Inverse Square,
Coming from the geometry, of spacetime curvature.

When his freshly minted law of warpage yielded,
Mercury's perihelion precession, Einstein was said
To have heart palpitations; he was beside himself with joy;
Three days he couldn't work; just so much to enjoy!

Einstein: "The years of searching in the dark for a truth
That one feels but cannot express, the intense desire and
The alternations of confidence and misgivings, until one
Breaks through to clarity and understanding, are known
Only to him, who has experienced them".[xxiii]

4.2 Bending of Path of Light

The bending of light by massive objects, Einstein predicted,
To finding it out, the Eddington expedition was dedicated;[xxiv]
At a solar eclipse, photographs of sun and stars were taken,
Behold, as Einstein said, starlight was bent by 1.75 seconds.

4.3 GPS Device

Seemingly impractical time curvature has a very practical app,
It's the ubiquitous GPS device, in phones, cars and laptops;
Without the tiny corrections to time curvature, due to gravity,
It's not possible to locate positions, with pin-point accuracy.

Part 5: Implications and Impacts

We turn now to reflect on relation, of two theories on gravity;
Here is an interesting case of two pictures, of the same reality;
The pictures are completely different from each other;
Is it reflecting us, reality, or something different altogether?

5.1 Comparison of Two Theories

5.1.1 Same Reality—Gravity: Two Pictures

Gravity is not a force;
Mass is its source;
Gravity is geometry;
Mass warps spacetime geometry.

What now we ask, "Is Einstein right"?
Is reality, geometry of spacetime curvatures?
Is his guess right?
Right, what is its source?

Yes, Einstein gives answers, where Newton had none;
Explains Mercury's orbit, bending of starlight near the sun;
Predicts black holes, from which even light cannot escape,
As well as frequency shift of light, falling into gravity's grip.

No, at black hole singularities, Einstein has no answers;[xxv]
No, it has no answer at Planck scale, at 1.62 X 10^{-35} meters;
Domain of very small and very massive is beyond its scope;
The Quantum and General Relativity have to fuse to cope.[91]

We have two completely different pictures of gravity;
One is force; the other, curvature of spacetime geometry;
Yet, each yields the same numbers within error limits,
In overlapping domain, where each an answer can generate.

There are two descriptions, when you jump out a window:
One says, force is pulling you down to ground below;
Other says, down a chute you're accelerating,
That spacetime curvature in vicinity is setting.

What is the reality, behind these two different pictures?
Is it force, or the geometry of spacetime curvatures?
This is a case where the fish you bag,
Depends on the kind of net you have.

Newton, with a simpler net: bagged reality as force;
Einstein, with a complex net: as spacetime curvature;
We have two descriptions, of the same underlying reality;
Gravity has two different images, yet no incompatibility.

The reality of gravity presents itself differently,
As force, when Newton with force hypothesis, made a query;
As geometry, when Einstein asked, with spacetime curvature;
The question one asks, sets the answer one gets—the picture.

---

[91] Quantum stands for Quantum Mechanics. It is a branch of Physics. Its principles govern the very small—elementary particles, atoms, molecules.

Einstein's more inclusive theory contains Newton's,
Einstein's equation reduces to Newton's, under conditions;
Mathematically, in their common domain, two are the same,
But what a difference in pictures, in psychology, each claims.

In terms of explanatory power, Einstein's is the greater;
"Action" at a distance makes Newton's the lesser;
Einstein's curvature of spacetime propagates at finite speed,
Like ripples on a pond's surface, touched by an insect behave.

5.1.2 Who is Right: Einstein or Newton? Domains of Validity

Where each gets numerical answers, in overlapping domain,
The mathematics of the two theories is the same;
This sameness reflects the "structure" of reality of gravity;
The difference in pictures stems from human psychology.

Can we say that Newton's theory is wrong?
A notion is prevalent, that a theory can only be proved wrong;
Can we say that Einstein's theory is right?
A notion is prevalent, that a theory can never be proved right.

The domain defined by speeds lower than the speed of light,
And gravity intensity below a threshold at a specified site;
Then, Newtonian calculations yield correct answers,
Despite the fact, it differs much in terms of pictures.

That Newton's theory, in its validity domain,
Has no chance of being shown wrong, will remain;
A theory that has stood the tests of time, with might;
Has every reason to claim, it has been proved right.

A skyscraper is eloquent witness, to Newton's rightness;
That it stands tall, testifies to his law's correctness;
The calculations, that went into making the building,
Can never be shown wrong, or found wanting.

To suggest designing a house, using General Relativity,
Betrays a lack of understanding, or downright stupidity;
Newton's theory is correct, within its validity domain,
Architecture and engineering, it'll continue to underpin.

5.1.3 Reality, Theory, Pictures and Mathematics

Domain defined: near speed of light, and near massive bodies,
Newton's theory is wrong; reality Einstein's theory embodies;
Mathematics of two theories is the same, in their intersection,
Thus, at its barest, Reality consists of mathematical relations.

A theory, then, consists of two components essential:
The picture, psychological; and the relations, mathematical;
In terms of picture, Newton's theory of gravitation is weak:
Force is instantaneous; has nothing its transmission to effect.

Einstein's theory is strong both in relations, and in picture,
Its mathematical relations will remain true, in the future;
Even as we're sure, that its picture will be rendered obsolete,
As soon as vaunted theory of Quantum Gravity is complete.

A picture of reality from a theory only lasts for so long,
Until a new picture, from the next theory comes along;
But, amply verified mathematical relations will remain,
So long are satisfied, the conditions of its validity domain.

5.1.4 Pyramid of Knowledge

Scientific knowledge is cumulative, like a pyramid;
Layer by layer, one on top of another we build;
An upper layer has a larger validity domain,
But, it does not invalidate a lower layer's gain.

5.2 Final Thoughts on Reality

Reality is not given, once and for all;
It is reached by action, which yields little by little;
To find the perfect human image, artists are ever chiselling;
To reach reality, scientists are ever experimenting, theorizing.

Reality is not a fixed entity, like a buried treasure;
It is more like a great piece of Art, or Literature;
In poems, poets attempt to capture human's core being;
In theories, scientists try to fathom "deep down things".[xxvi]

Theories of Newton and Einstein, above ideas, demonstrate;
At domain where both have validity, "Reality" equivocates:
As Newton probed its form, it showed the "shape" of force;
As Einstein queried, its "form" was spacetime curvature.

What is it? Is gravity, force or spacetime curvature?
In the domain where both are valid, at their core,
Mathematically, the two theories are identical;
Reality of gravity then, is mathematical.

At bottom, Einstein's and Newton's theories on gravity,
Affirm Galileo's stance: parts of reality relate mathematically;
At bottom, Nature seems not to care for the picture;
However, our minds need a crutch, like force or curvature.

5.3 Mind and Nature

Black holes, neutron stars, white dwarfs[92]
Seem to pay attention to spacetime warps,
Contained in the field equation of Einstein;
Is it a pre-established harmony, of Nature and mind?

Mind evolved from Nature;
In its architecture,
Are imprinted Nature's laws;
This, our thinkers amply show.

We all are expressions, of Nature's Architecture;
It must be no surprise then, Einstein's thoughts mirror,
The reality of what is going on "deep down things";
It is a case of two mirrors: one, the other mirroring.

We are designed to understand the Cosmos;
We are designed to search for Nature's laws;
Our sense of being human is tied to Nature;
Could it be otherwise? We take after our Mother.

What does this imply?
Children of Nature that we are, we can't deny
The call of Nature, its laws to find;
It is the same as the craving of our mind.

---

[92] Black hole is a warping of a spacetime region in which nothing can
escape, including light. Neutron stars are made up of neutrons,
remnants of a gravitational collapse of a massive star. White dwarf is a
stellar remnant mainly composed of electron-degenerate matter. The
electrons are stripped from their nuclei and are highly compressed.

It is a plea from Mother;
A plea the child is attuned to hear;
To heed the plea is HUMANITY'S highest calling;
This is what is to be FULLY, a HUMAN BEING.

If I may, not only is Science relevant,
In satisfying our material needs and wants;
But, it also offers insights, into the human condition:
WHAT IT IS TO BE HUMAN, the perennial question.

Meaning; what it is to be human, rest on who we are:
We all are embodiments, of the laws of Nature;
In our bodies, are sculpted the laws of Nature;
In our minds, are sculpted the laws of Nature.

The search for the laws of Nature,
Is a search for **who we are**,
Our origins, our roots; in our chase,
We find reflections of our "face".

To know the Mind of Nature,
Is to know the Human Core;
To appreciate Nature's Beauty,
Is at the heart of our Humanity.

5.4 No Authority in Science

To say what is right, there's no authority in Science whatsoever;
A method that critiques ideas with evidence is all what is there;
By which we muster resources to show where the idea is wrong;
It is the reason progress in future will continue to grow strong.

5.5 Science: A Continuing International Enterprise

To Newton, we give thanks for pioneering Nature's study,
Systematically using experiment, mathematics and theory;
To Einstein, we give thanks for his childlike greatest insight:
Curvature of spacetime is gravity, we learn with great delight.

And now, we bring to a close, our discussion of gravity;
Gravity's riddle has been with us, as long as our history;
Two giant strides we have taken, in understanding gravitation,
With iconic Albert Einstein, and magical Sir Isaac Newton.

Even as we draw the curtain, the quest for gravity goes on;
Of Quantum-General Relativity, we have yet to make a fusion;
Results from LHC will surely make contributions,[93]
Toward elucidating the continuing riddle of gravitation.

---

[93] LHC stands for Large Hadron Collider, the biggest particle accelerator in the world.

Conclusion

We have come to the end; a start of a new beginning;
At introduction of Book I, we looked at the historical setting;
We reviewed our explorations, of our cosmic Residence;
Here, we reflect on meaning, the place of **homo sapiens**.

I will draw, not only from Physics: its implications,
But also, I will draw equally from Biology: its lessons;
Even among the Sciences, there is a reluctance to straddle,
I appreciate the courage, to brave accusations of "twaddle".

What is the present cosmic Architecture?
What is our place in the scheme of Nature?
What does it mean to be human?
What can we draw from the Einstein "phenomenon"?

Present Cosmic Architecture

We have a single, one-tiered, unified Architecture;
The same laws of Physics apply anytime, anywhere;
The same constituents compose all things, living and non-living;
The same mechanisms govern all things, living and non-living.

That is the Architecture, of the Residence of *homo sapiens;*
It's yet unfinished; efforts on-going would clarify our origins;
The very small and the very large: we have yet to unify;
Some think it is imminent; others, it may take a long while.

What it means to be Human

We all are sons and daughters, of Nature;
We all are expressions, of Nature's Architecture;
We all are embodiments, of the laws of Nature;
We all are made of the same stuff, from dying stars.

In our bodies, are sculpted the laws of Nature;
In our minds, are sculpted the laws of Nature;
We all have the same kinship, with all living things;
We all share the same atoms, with all non-living things.

Whether we are religious, or not;
Whether we are believers, or not;
It is our duty, to know the basic facts of our existence;
Facts, of who we are, must not be a matter of expedience.

Facts, of who we are, must be just plain matters of fact;
The facts enumerated above, are just plainly that;
With above enumerated facts, if we have trouble,
We have to re-examine assumptions, soonest practicable.

To be human, is to feel a bond of kinship, with all living:
From amoeba, plants, birds, fish, whales, to rodents;
To be human, is to feel a bond of unity, with all non-living:
From quarks, light photons, to periodic table elements.

Controversial: to be human, Is to see our face, in chimps;
This means, we acknowledge our "animal" origins;
This does not imply, that we condone "animality" in humans;
From "brute" to human beings, we recognize the advance.

To be human, is to use the mirror of Nature in our brains;
Thru this mirror, we view, understand Nature's occurrences;
The mirrors are of varying clarity; Einstein's was the clearest;
In some brains, the mirror is not whole but broken to pieces.

To be human, is to recognize our Mother;
To be human, is to accept who we are;
To deny the basic facts of existence is to deny our Mother;
To reject the basic facts of existence is to reject who we are.

Our Place in Nature

Our place in Nature is not "special";
Our hold on existence is ever fragile;
Our present domination is not inevitable;
To a rogue asteroid, our existence is vulnerable.

To appreciate the contingency, of our existence,
Imagine the Evolution Clock, is set to re-commence;
Humans are very highly unlikely, to be in the repeat;
The space of possible life forms is infinite.

Our species is just one among millions;
If we include extinct ones, one among billions;
If we feel the inevitability of humans' rise, dominance,
Remember, species survival rests purely on chance.

"Special" has no place, in the ways of Nature;
To the Eye of Chance, every species' tenure is unsecure;
Lady Luck does not favour, this species or that species;
She just continues to roll the dice, in her relentless ways.

"Privilege" has no place, in the doings of Nature;
From no privileged position in Cosmos: no Centre;
To no privileged frame, in the workings of laws of Physics;
And in the evolution of life, there is no privileged species.

Our species has come, to its present pre-eminence;
Not by its merits; but by the grace of almighty Chance;
What comes by chance, by chance it will be gone;
Such is the fragility, of our species' existence.

We share this Cosmos, with fellow earth voyagers;
We must act with ears, attuned to past messengers;
Our present supremacy is as precarious, as a coin flip;
More than once in our past, to oblivion we almost slipped.

Adapting the equivalence of inertial frames,
We enunciate this principle: no species is privileged, all species
Are equal or equivalent; of course, obviously humans,
Dominate planet Earth; and ever since have been dominant.

My point is that awareness of our precariousness,
Coupled with the principle of equivalence of species;
Might just possibly temper the rapaciousness,
With which we exploit the common earth resources.

Our place in the Cosmos is not one, of mastership;
But, it is one of responsible stewardship;
Guided by the tenuous hold, we have on our existence,
By species equivalence, we aim for: co-equal, co-existence.

Phenomenon of Einstein

Mass and pressure tell spacetime how to curve,
Spacetime tells masses how to move;
In place of forces' pushes and pulls,
Is the geometry of spacetime curvatures.

The above, in brief, is the cosmic blueprint;
It came out of Einstein's brain;
But, where it came from, is an important question;
This brings us, to the Einstein "phenomenon".

By the Einstein phenomenon, I mean the clear,
Correspondence between his thoughts and Nature;
It is true for other thinkers too; but, Einstein stands out;
Of this observation, what do we make out?

Being of Nature, some of us have brains that match Nature's;
To name a few: Galileo, Newton, Maxwell, Kepler;
Reasoning from the principle of adequation, it must be,
That some brains reflect some aspects of reality.

I would go further than this: all humans have a copy,
Of Nature's blueprint in our brains; only they vary in clarity;
This is what enables us to understand the print,
Out of the great minds: like that from Einstein's brain.

Every time we read a book or think,
We are tracing sections, of the cosmic blueprint;
We need the aid of printouts, from the great minds;
They enable us to trace the blueprint in our minds.

Once in a long while: like every two to three hundred years,
An unusual brain comes, with a clearer mirror of Nature;
My neck is out: the brain that fuses Quantum & Gravity,
Is yet to be born, about the end of this twenty-first century!

"I want to know God's thoughts. The rest are details." Einstein;
I take the quote above, as indicative of Einstein's unusual mind;
Such a brain is needed, to fuse Quantum and General Relativity;
It's the next giant step, in unfolding the Architecture of Reality.

Epilogue

Fragmentation seems to be a price we pay for "progress". In areas of human knowledge, it seems depth is priced much higher than breadth. In consequence, the race has been going on to be the expert on the "narrowest" area possible. Incentives have been so awarded accordingly. Someone who straddles a wider area is viewed as dilettante at best, or with contempt at worst.

A theme running through both books is bridging the divide between the Arts and the Sciences. The age-old question is: *how can the Two ever meet?* On one hand, the Arts directly deal with aspects of Human existence. On the other, the Sciences directly deal with "inanimate" matter. Their subjects could not be more different. Only through what I call the "humanizing" aspects of Science can the Two ever meet: the implications of the findings of Science, the insights from Science, on the *human condition*.

Thus, the Two can meet on the ***illuminations*** each brings to the perennial question: *WHAT IT IS TO BE HUMAN.* This is the intersection between the two areas of Human Knowledge; the zone of common concern between the two Cultures.

Since the early humans, we have been exploring the "inanimate" till the present. Along the way, we have learned a lot. In the end, it turns out that the outward journey of our exploring is actually an inward exploration into our being. The sum of all our knowledge of the "inanimate" points to one direction: understanding ***who we are.*** Meaning of life and what it is to be human, depend on *who we are.*

In all our exploring, we come to this realization: *we are expressions of Nature's Architecture. We are embodiments of the laws of Nature. The search for laws of Nature is a search for who we are.* This is the answer from Science to the question: *what it is to be human.*

The poet, T S Eliot, expressed our exploring most beautifully:[94]

We shall not cease from exploration
And the end of all our exploring
Will be to arrive where we started
And know the place for the first time.

---

[94] The quote is from "Little Gidding" (No 4 of 'Four Quartets') by T S Eliot.

---

[i] Whenever you encounter "reference frame", or just "frame", or "inertial frame"—imagine space and time in that frame filled with clocks and meter sticks forming a lattice. (Taylor & Wheeler, 1963)

[ii] The quote is taken from "Essay on Criticism", by Alexander Pope (Price, 1970).

[iii] Lorentz Transformation:

In 2-dimensional spacetime, we have time, the vertical axis, and one dimension, the horizontal axis. Let (z_0, z_1) be Chu's coordinates and (x_0, x_1) be Sue's coordinates, where subscript 0 refers to time, and subscript 1 to one dimension of space. For example, z_0 is the time coordinate of Chu; while, x_0 is the time coordinate of Sue. These coordinates, of Sue and Chu, are related by a Lorentz transformation, a set of equations which can be derived from the postulates of special relativity; (note speed of light is one)

1) $x_0 = 1/\text{sqrt} (1-v^2) * (z_0 + v * z_1)$
2) $x_1 = 1/\text{sqrt} (1-v^2) * (z_1 + v * z_0)$

which holds for any spacetime point, i.e., in Sue's coordinates, Chu is moving with speed v along positive x_1 direction.

For Chu, the coordinates of event A: (0, 0);
the coordinates of event B: $(2L_0, 0)$.

For Sue, from Lorentz transformation,
the coordinates of event A: (0, 0),
as shown in calculation below:

$x_0 = 1/\text{sqrt} (1-v^2) * (z_0 + v * z_1) = 1/\text{sqrt} (1-v^2) * (0 + v * 0) = 0.$
$x_1 = 1/\text{sqrt} (1-v^2) * (z_1 + v * z_0) = 1/\text{sqrt} (1-v^2) * (0 + v * 0) = 0.$

For Sue, the coordinates of event B: $2L_0/\text{sqrt} (1- v^2) * (1, v)$
as shown in calculation below:

$x_0 = 1/\text{sqrt} (1-v^2) * (z_0 + v * z_1) = 1/\text{sqrt} (1-v^2) * (2L_0 + v * 0)$
$= 2L_0/\text{sqrt} (1-v^2)$
$x_1 = 1/\text{sqrt} (1-v^2) * (z_1 + v * z_0) = 1/\text{sqrt} (1-v^2) * (0 + v * 2L_0)$
$= 2L_0/\text{sqrt} (1-v^2) *v.$

We see that for Chu, $(time)^2 - (space)^2 = (2L_0)^2$.
For Sue, we have

$$(time)^2 - (space)^2 = \{ 2L_0/\text{sqrt} (1-v^2) \}^2 - \{ 2L_0/\text{sqrt} (1-v^2) * v \}^2$$
$$= (2L_0)^2 * \{ 1/(1-v^2) - v^2 * 1/(1-v^2) \}$$
$$= (2L_0)^2 * (1-v^2)/(1-v^2)$$
$$= (2L_0)^2$$

Thus, the interval is invariant.

[iv] In the same way that Sue concludes that Chu's clock runs more slowly than hers, we will show that Chu also concludes Sue's clocks are running more slowly than his. From the Lorentz transformation we performed, Sue has distance $D = 2L_0/\text{sqrt}(1-v^2) * v$, and time $T = 2L_0/\text{sqrt}(1-v^2)$ between events A and B. For Chu, time $= 2L_0$ and distance 0 between the same two events. Sue claims that clock at A and at B are synchronized. We know from Part 5, Book I, that simultaneity is relative; it depends on the frame. To Chu, who sees Sue frame is moving to left at speed v toward event A, the two clocks at rest in Sue frame are not synchronous: they do not show the same time.

Now, let us determine the time difference between the two clocks which are not synchronous according to Chu. Light flash from midpoint will reach clock B sooner than at clock A. Remember the speed of light is one. Time t_B is the time when flash reaches clock B. To Chu, distance is length contracted, $D \text{ sqrt}(1-v^2)$. So, $D \text{ sqrt}(1-v^2)/2 - v t_B = 1*t_B$. Solving for $t_B = D \text{ sqrt}(1-v^2)/2 * 1/(1+v)$. Similarly, $t_A = D \text{ sqrt}(1-v^2)/2 * 1/(1-v)$. The time difference between the starting of the two clocks is

$$t_{AB} = t_A - t_B = D \text{ sqrt}(1-v^2)/2 * [1/(1-v) - 1/(1+v)]$$
$$= D \text{ sqrt}(1-v^2)/2 *[2v/(1 - v^2)]$$
$$= D v * 1/\text{sqrt}(1-v^2).$$

This is the time difference t_{AB} between the start of clock B to the start of clock A. But Chu further notices that both clocks run more slowly by the shrink factor: $\text{sqrt}(1-v^2)$. In effect, Chu finds clock B has ticked

$$\text{sqrt}(1-v^2) * D v * 1/\text{sqrt}(1-v^2) = D * v \text{ seconds}$$

just before clock A starts. So we have established that clock A lags behind clock B by D * v, from the point of view of Chu. Now, we retrace the motion of Chu. Chu starts from event A, where he synchronizes his clock with Sue's clock A: both read 0; and moves towards B at speed v. When Chu reaches B, Sue's clock at B reads $2L_0/sqrt(1-v^2)$, while Chu's clock reads

time in moving clock = shrink factor * time in stationary clock
= $sqrt(1-v^2) * 2L_0/sqrt(1-v^2)$
= $2L_0$.

Now, how can Chu claim that the clocks of Sue run more slowly than his? By the non-synchronicity of the clocks in Sue frame. We have shown that clock A lags behind Clock B by D * v. Taking account of the lag of clock A, to get the reading of clock A, he subtracts from clock B reading the lag time of clock A:

time in clock A
= (time in B clock) – (lag time of A clock)
= $2L_0/sqrt(1-v^2) - D *v$
= $2L_0/sqrt(1-v^2) - 2L_0/sqrt(1-v^2) *v *v$
= $2L_0/sqrt(1-v^2) [1-v^2]$
= $sqrt(1-v^2) * 2L_0$.

So clocks of Sue run more slowly and have just the right shrink factor with respect to Chu's clock time! The claim is based on the relativity of simultaneity!!

[v] The derivation of the metric equation follows. Sue measures the distance between A and B to be D; and time to be T. The total zigzag path of light beam is $2* sqrt \{ (L_0)^2 + (D/2)^2 \}$. Since the speed of light is 1, time T is equal to total path. So, $T = 2 * sqrt \{ (L_0)^2 + (D/2)^2 \} = sqrt \{ (2L_0)^2 + (D)^2 \}$. The interval, **S**, is defined as the time a single clock moving at constant velocity takes to move from event A to B. In our case, $2L_0 = $ **S**. After substitution and algebraic manipulation, we have the so-called metric equation: $S^2 = T^2 – D^2$. In words, Interval square equals Time square minus Distance square.

This is the case for time-like related events, where time separation is greater than space separation. As we will see in section 6.6, Book I, there are two other kinds of relation between events. Space-like related events, the interval is written: $S^2 = D^2 - T^2$, or Interval square equals Distance square minus Time square. The last is the light-like related events, where Interval is zero, because the time separation is equal to space separation. These are events that occur to or set off by light. (Moore, 1995).

[vi] Time dilation derivation follows. We factor out T and take the square root, we get $S = T * sqrt \{ 1 - (D/T)^2 \}$. The ratio D/T is just the velocity, v, of Chu frame. $S = T * sqrt \{ 1 - v^2 \}$. In this equation, we reinterpret S to generalize it: S is the time any single clock registers however it moves from one event to another. We call it the "proper time". When the clock moves at constant velocity between events, then it is equal to the interval. At any other velocity, it is proper time; we denote it by the symbol—T_0. The time T, on the other hand, is called coordinate time. It is time as measured by two stationary clocks, one at event A and the other at event B. So, $T_0 = T * sqrt \{ 1 - v^2 \}$. The slowing down factor or the shrink factor is $sqrt \{1 - v^2\}$. So, a clock moving relative to another clock slows down (Moore, 1995). [Strictly speaking, I should show that for very short path, proper time is about equal to the interval. In the limit of infinitesimal path, proper time equals the interval].

[vii] From time dilation equation, we have $T_0 = T * sqrt \{ 1 - v^2 \}$, where T_0 corresponds to age of Alice and T to age of Sue. We know from what is given in the problem that T_0 is 14. T, age of Sue, is given by $T = T_0 / sqrt \{ 1 - v^2 \} = 14/ sqrt \{ 1 - (0.96)^2 \} = 50$.

[viii] The total speed equation is the time dilation re-arranged, as given in endnote vi. We have $T_0 = T * sqrt \{1 - v^2\}$, where T_0 is the time in clock in a moving frame, T is the time in clock in a stationary frame, and v is ordinary speed distance/time. Dividing the equation by T, then squaring the result, and finally grouping terms, we get $(T_0/T)^2 + v^2 = 1$. With our choice of units (expressing speeds as fractions of speed of light), the speed of light is one. The equation in words: speed thru time square + speed thru space square = speed of light square. The ratio T_0/T is the speed of object thru time (which is the same as the shrink factor) because the ratio is a measure of how fast time in clock in the moving frame is slowing down compared to time in clock in the stationary frame. The v is the speed thru space as a fraction of the speed of light. There is, therefore, a negative relation between the

speed thru space v and the speed thru time (T_0/T). This is really just a different presentation of the time dilation equation. The equation says the combined speed of any object equals the speed of light. We all moved at light speed!!!

[ix] Length contraction derivation follows. At time t_1, the distance light beam travels is $L + v t_1 = 1* t_1$. Here 1 is the speed of light and speed v is expressed as a fraction of light speed. Time t_1 is $t_1 = L/(1 - v)$. At time t_2, the distance light beam travels is $L - v t_2 = 1* t_2$. Time t_2 is $t_2 = L/(1 + v)$. Adding t_1 and t_2, we obtain $t_1 + t_2 = T = 2L/(1 - v^2)$. This T is the same coordinate time T we had in the formula for time dilation: $S = T * sqrt \{1 - v^2\}$. Remember, the light clock, whose length Sue is measuring, is lying horizontal in Chu frame. Solving for T and substituting $2L_0$ for S, we get $T = 2L_0/sqrt \{1 - v^2\}$. Comparing the T expressions, we get $L = L_0 * sqrt \{1 - v^2\}$. Stick in relative motion has length L equal to length in stationary frame multiplied by the shrink factor. A stick in relative motion shrinks (French, 1968).

[x] The phrase in quotes, "deep down things" is taken from the poem "God's Grandeur" by Gerard Manly Hopkins: "And for all this, nature is never spent; There lives the dearest freshness deep down things" (Mack, Dean, & Frost, 1950).

[xi] There is a difference, one part in a billion, in the distance between sun and mercury, depending on the body you measure the distance from: Mercury or Sun. From a fundamental point of view, it should not matter from which body one measures the distance. (Thorne, 1994)

[xii] Inertial frame is a lattice of clocks and sticks moving at constant speed and direction (Taylor & Wheeler, 1963).

[xiii] If you want to exercise your imagination, this is a problem most apt. We begin by considering the vertical direction.

One, you and opaque screen with an adjustable horizontal slot move as one.

Two, the ball, you and screen start at the same time and are in free fall throughout the period—all have the same downward acceleration and thus the same change in velocity downward. While ball starts with initial velocity up, you have zero initial velocity—up or down.

Three, the ball's initial velocity is changing by a negative amount equal to that of you-screen, until the ball reaches its peak, at which point its velocity is momentarily zero. When ball is at peak, you and screen pick up downward speed equal to initial speed of ball up.

Four, in the first phase of motion, ball goes up, while you and screen go down. In the second phase, all are going down. At start of second phase, ball has zero initial speed down, while you-screen have initial speed down equal to initial speed of ball up.

Five, vertical velocity of ball has two components: constant initial up from the toy gun and a velocity down changing with time due to gravity. You-screen have no initial velocity; only velocity down changing with time due to gravity. The equations are: $Y(ball) = V_o\, t - \frac{1}{2} g\, t^2$; $Y(you\text{-}screen) = -\frac{1}{2} g\, t^2$, where Y is height with origin at gun level; V_o is initial velocity up of ball; t is time.

Six, as long as the ball is going up to its peak, you have to move the slot up equal to the constant initial velocity up of the ball— in order to see the ball. The only difference between the ball and you-screen is the initial velocity up of the ball. So, the horizontal slot of screen has the equation: $Y(horizontal\ slot) = V_o\, t - \frac{1}{2} g\, t^2$, just like the ball.

Seven, in the downward phase, you have to move the slot up at the same speed up as before to see the ball all the time. Since the horizontal slot is part of you-screen, you have to add a velocity up to the terms of you-screen and thus, resulting in the following equations: $Y(ball) = -\frac{1}{2} g\, t^2$; $Y(you\text{-}screen) = -V_o\, t - \frac{1}{2} g\, t^2$ $Y(horizontal\ slot) = V_o\, t + Y(you\text{-}screen) = V_o\, t - V_o\, t - \frac{1}{2} g\, t^2 = -\frac{1}{2} g\, t^2$. Like in the up episode, the only difference between the ball and you-screen is the initial velocity down of you-screen. Note, the ball and the horizontal slot have the same equation.

Now, the horizontal direction: The ball has a constant horizontal velocity component. So, you see the ball through the slot, moving straight with a tilt to the right.

Altogether, in your free-fall frame, you see the ball moving in a straight line at constant speed, tilted to the right. (Thorne, 1994).

[xiv] The elevator is accelerating upward. This means its speed is increasing all the time, which in turn means that it covers the same distance in ever shorter time. As a result, the T clock's signal, located at the top of the elevator, every second it ticks, arrives at the floor where the D clock is located, faster and faster. So, the T clock at the top ticks faster than the D clock down the floor. By the Equivalence Principle, you can't distinguish this effect in an accelerating elevator

from the effects of gravity. Therefore, mass distorts time in different places: the farther from the mass (which is equivalent to "the farther from the floor, the faster the ticking rate"), the faster the clock ticks. This is what is meant by time curvature (Feynman, Leighton, & Sands, 1964).

[xv] This is "radius" at which any object including light cannot escape from being swallowed (Thorne, 1994).

[xvi] Kip Thorne's thought experiment as found in (Thorne, 1994).

[xvii] Special Relativity shows that clocks, in relative motion, measure different time durations. The ticking rate for a clock moving relative to another is slower. The rate depends on the square of its speed. For a clock to show the longest time, it should not move at all. To show the shortest time, the clock should move the fastest (Greene, 2004).

[xviii] The quote from Kip Thorne "Black Holes and Time Warps" (Thorne, 1994)

[xix] Source is "Black Holes and Time Warps" by Kip Thorne. (Thorne, 1994)

[xx] Ordinarily, mass is the sole determinant of spacetime curvature; except in a few exotic places, like in neutron stars, where pressure contributes significantly (Thorne, 1994).

[xxi] I modified the statement found in the book Gravitation by Misner, Thorne and Wheeler (Misner, Thorne, & Wheeler, 1973).

[xxii] One of the puzzles related to planetary orbits is the advance of the perihelion of Mercury, the point closest to the sun. As Mercury revolves around the sun, it does not trace the same ellipse, such that its perihelion advances around the sun. Much of the advance was accounted for by the gravitational pull of the other planets. But a minute amount of 43 arc seconds per century cannot be accounted for by Newton's theory. But, its explanation by curvature of spacetime set Einstein heart to flutter, when his calculation showed 43 arc seconds. (Thorne, 1994)

[xxiii] The Einstein quote from (Thorne, 1994).

[xxiv] In 1919, an expedition was organized under the leadership of Sir Arthur Eddington, the Astronomer Royal of England, with the express purpose of measuring the deflection of light from stars as they graze

the sun, during the solar eclipse. The measurements confirmed the prediction of Einstein: 1.75 seconds of arc deflection.

[xxv] Black hole is formed when the mass of an object exceeds a certain number of solar masses. When its nuclear fuel is exhausted, its gravitational fields will have no countervailing force and are so strong that the object continues to contract where all the mass is compressed to a point. This point of enormous density is a singularity.

[xxvi] The phrase in quotes, "deep down things" is taken from the poem "God's Grandeur" by Gerard Manly Hopkins: "And for all this, nature is never spent; There lives the dearest freshness deep down things" (Mack, Dean, & Frost, 1950)

References:

Drout, P. M. (2009). *A Way With Words IV: Understanding Poetry.* Recorded Books.

Feynman, R. P., Leighton, R. B., & Sands, M. (1964). *The Feynman Lectures in Physics* (Vol. II). Menlo Park: Adison-Wesley Publishing Company.

Greene, B. (2004). *The Fabric of the Cosmos.* London: Penguin Group .

Mack, M., Dean, L., & Frost, W. (Eds.). (1950). *Modern Poetry* (Vol. VII). Englewood Cliffs: Prentice - Hall, Inc.

Misner, C. W., Thorne, K. S., & Wheeler, J. A. (1973). *Gravitation.* San Francisco: W. H. Freeman.

Taylor, E. F., & Wheeler, J. A. (1963). *Spacetime Physics.* San Francisco: W. H. Freeman.

Thorne, K. S. (1994). *Black Holes and Time Warps: Einstein's Outrageous Legacy.* London: Papermac.